Approximation Algorithms and Semidefinite Programming

Bernd Gärtner • Jiří Matoušek

Approximation Algorithms and Semidefinite Programming

 Springer

Bernd Gärtner
ETH Zurich
Institute of Theoretical Computer Science
8092 Zurich
Switzerland
gaertner@inf.ethz.ch

Jiří Matoušek
Charles University
Department of Applied Mathematics
Malostranské nám. 25
118 00 Prague 1
Czech Republic
matousek@kam.mff.cuni.cz

ISBN 978-3-642-43332-0   ISBN 978-3-642-22015-9 (eBook)
DOI 10.1007/978-3-642-22015-9
Springer Heidelberg Dordrecht London New York

Mathematics Subject Classification (2010): 68W25, 90C22

© Springer-Verlag Berlin Heidelberg 2012
Softcover reprint of the hardcover 1st edition 2012

Printed on acid-free paper

Springer is part of Springer Science+Business Media (www.springer.com)

Preface

This text, based on a graduate course taught by the authors, introduces the reader to selected aspects of semidefinite programming and its use in approximation algorithms. It covers the basics as well as a significant amount of recent and more advanced material, sometimes on the edge of current research.

Methods based on semidefinite programming have been the big thing in optimization since the 1990s, just as methods based on linear programming had been the big thing before that – at least this seems to be a reasonable picture from the point of view of a computer scientist. Semidefinite programs constitute one of the largest classes of optimization problems that can be solved reasonably efficiently – both in theory and in practice. They play an important role in a variety of research areas, such as combinatorial optimization, approximation algorithms, computational complexity, graph theory, geometry, real algebraic geometry, and quantum computing.

We develop the basic theory of semidefinite programming; we present one of the known efficient algorithms in detail, and we describe the principles of some others. As for applications, we focus on approximation algorithms.

There are many important computational problems, such as MaxCut,[1] for which one cannot expect to obtain an exact solution efficiently, and in such cases one has to settle for approximate solutions.

The main theoretical goal in this situation is to find efficient (polynomial-time) algorithms that always compute an approximate solution of some guaranteed quality. For example, if an algorithm returns, for every possible input, a solution whose quality is at least 87% of the optimum, we say that such an algorithm has *approximation ratio* 0.87.

In the early 1990s it was understood that for MaxCut and several other problems, a method based on semidefinite programming yields a better approximation ratio than any other known approach. But the question

---

[1] Dividing the vertex set of a graph into two parts interconnected by as many edges as possible.

remained, could this approximation ratio be further improved, perhaps by some new method?

For several important computational problems, a similar question was solved in an amazing wave of progress, also in the early 1990s: the best approximation ratio attainable by *any* polynomial-time algorithm (assuming $P \neq NP$) was determined precisely in these cases.

For MAXCUT and its relatives, a tentative but fascinating answer came considerably later. It tells us that the algorithms based on semidefinite programming deliver the best possible approximation ratio, among all possible polynomial-time algorithms. It is tentative since it relies on an unproven (but appealing) conjecture, the Unique Games Conjecture (UGC). But if one believes in that conjecture, then semidefinite programming is the ultimate tool for these problems – no other method, known or yet to be discovered, can bring us any further.

We will follow the "semidefinite side" of these developments, presenting some of the main ideas behind approximation algorithms based on semidefinite programming.

The origins of this book. When we wrote a thin book on linear programming some years ago, Nati Linial told us that we should include semidefinite programming as well. For various reasons we did not, but since one should trust Nati's fantastic instinct for what is, or will become, important in theoretical computer science, we have kept that suggestion in mind.

In 2008, also motivated by the stunning progress in the field, we decided to give a course on the topics of the present book at ETH Zurich. So we came to the question, *what* should we teach in a one-semester course? Somewhat naively, we imagined we could more or less use some standard text, perhaps with a few additions of recent results.

To make a long story short, we have not found any directly teachable text, standard or not, that would cover a significant part of our intended scope. So we ended up reading stacks of research papers, producing detailed lecture notes, and later reworking and publishing them. This book is the result.

Some FAQs. Q: *Why are there two parts that look so different in typography and style?*

A: Each of the authors wrote one of the parts in his own style. We have not seen sufficiently compelling reasons for trying to unify the style. Also see the next answer.

Q: *Why does the second part have this strange itemized format – is it just some kind of a draft?*

A: It is not a draft; it has been proofread and polished about as much as other books of the second author. The unusual form is intentional; the (experimental) idea is to split the material into small and hierarchically organized chunks of text. This is based on the author's own experience with learning things, as well as on observing how others work with textbooks. It should make the text easier to digest (for many people at least) and to memorize the most important things. It probably reads more slowly, but it is also more compact than a traditional text. The top-level items are systematically numbered for an easy reference. Of course, the readers are invited to form their own opinion on the suitability of such a presentation.

Q: *Why haven't you included many more references and historical remarks?*

A: Our primary goal is to communicate the key ideas. One usually does not provide the students with many references in class, and adding survey-style references would change the character of the book. Several surveys are available, and readers who need more detailed references or a better overview of known results on a particular topic should have no great problems looking them up given the modern technology.

Q: *Why don't you cover more about the Unique Games Conjecture and inap-proximability, which seems to be one of the main and most exciting research directions in approximation algorithms?*

A: Our main focus is the use of semidefinite programming, while the UGC concerns lower bounds (inapproximability). We do introduce the conjecture and cite results derived from it, but we have decided not to go into the technical machinery around it, mainly because this would probably double the current size of the book.

Q: *Why is topic X not covered? How did you select the material?*

A: We mainly wanted to build a reasonable course that could be taught in one semester. In the current flood of information, we believe that less mate-rial is often better than more. We have tried to select results that we perceive as significant, beautiful, and technically manageable for class presentation. One of our criteria was also the possibility of demonstrating various general methods of mathematics and computer science in action on concrete exam-ples.

Sources. As basic sources of information on semidefinite programming in general one can use the *Handbook of Semidefinite Programming* [WSV00] and the surveys by Laurent and Rendl [LR05] and Vandenberghe and Boyd [VB96]. There is also a brand new handbook in the making [AL11]. The books by Ben-Tal and Nemirovski [BTN01] and by Boyd and Vandenberghe [BV04] are excellent sources as well, with a somewhat wider scope. The lecture notes by Ye [Ye04] may also develop into a book in the near future.

A new extensive monograph on approximation algorithms, including a significant amount of material on semidefinite programming, has recently been completed by Williamson and Shmoys [WS11]. Another source worth mentioning are Lovász' lecture notes on semidefinite programming [Lov03], beautiful as usual but not including recent results.

Lots of excellent material can be found in the transient world of the Internet, often in the form of slides or course notes. A site devoted to semidefinite programming is maintained by Helmberg

[Hel10], and another current site full of interesting resources is `http://homepages.cwi.nl/~monique/ow-seminar-sdp/` by Laurent. We have particularly benefited from slides by Arora (`http://pikomat.mff.cuni.cz/honza/napio/arora.pdf`), by Feige (`http://www.wisdom.weizmann.ac.il/~feige/Slides/sdpslides.ppt`), by Zwick (`www.cs.tau.ac.il/~zwick/slides/SDP-UKCRC.ppt`), and by Raghavendra (several sets at `http://www.cc.gatech.edu/fac/praghave/`). A transient world indeed – some of the materials we found while preparing the course in 2009 were no longer on-line in mid-2010.

For recent results around the UGC and inapproximability, one of the best sources known to us is Raghavendra's thesis [Rag09]. The DIMACS lecture notes [HCA$^+$10] (with 17 authors!) appeared only after our book was nearly finished, and so did two nice surveys by Khot [Kho10a, Kho10b].

In another direction, the lecture notes by Vallentin [Val08] present interactions of semidefinite programming with harmonic analysis, resulting in remarkable outcomes. Very enlightening course notes by Parrilo [Par06] treat the use of semidefinite programming in the optimization of multivariate polynomials and such. A recent book by Lasserre [Las10] also covers this kind of topics.

Prerequisites. We assume basic knowledge of mathematics from standard undergraduate curricula; most often we make use of linear algebra and basic notions of graph theory. We also expect a certain degree of mathematical maturity, e.g., the ability to fill in routine details in calculations or in proofs. Finally, we do not spend much time on motivation, such as why it is interesting and important to be able to compute good graph colorings – in this respect, we also rely on the reader's previous education.

Acknowledgments. We would like to thank Sanjeev Arora, Michel Baes, Nikhil Bansal, Elad Hazan, Martin Jaggi, Nati Linial, Prasad Raghavendra, Tamás Terlaky, Dominik Scheder, and Yinyu Ye for useful comments, suggestions, materials, etc., Helena Nyklová for a great help with typesetting, and Ruth Allewelt, Ute McCrory, and Martin Peters from Springer Heidelberg for a perfect collaboration (as usual).

Errors. If you find errors in the book, especially serious ones, we would appreciate it if you would let us know (email: `matousek@kam.mff.cuni.cz`, `gaertner@inf.ethz.ch`). We plan to post a list of errors at `http://www.inf.ethz.ch/personal/gaertner/sdpbook`.

Contents

Part I
(by Bernd Gärtner)

Chapter 1
Introduction: MAXCUT Via Semidefinite Programming

Semidefinite programming is considered among the most powerful tools in the theory and practice of approximation algorithms. We begin our exposition with the Goemans–Williamson algorithm for the MAXCUT problem (i.e., the problem of computing an edge cut with the maximum possible number of edges in a given graph). This is the first approximation algorithm (from 1995) based on semidefinite programming and it still belongs among the simplest and most impressive results in this area.

However, it should be said that semidefinite programming entered the field of combinatorial optimization considerably earlier, through a fundamental 1979 paper of Lovász [Lov79], in which he introduced the *theta function* of a graph. This is a somewhat more advanced concept, which we will encounter later on.

In this chapter we focus on the Goemans–Williamson algorithm, while semidefinite programming is used as a black box. In the next chapter we will start discussing it in more detail.

1.1 The MAXCUT Problem

MAXCUT is the following computational problem: We are given a graph $G = (V, E)$ as the input, and we want to find a partition of the vertex set into two subsets, S and its complement $V \setminus S$, such that the number of edges going between S and $V \setminus S$ is maximized.

More formally, we define a *cut* in a graph $G = (V, E)$ as a pair $(S, V \setminus S)$, where $S \subseteq V$. The *edge set* of the cut $(S, V \setminus S)$ is

$$E(S, V \setminus S) = \{e \in E : |e \cap S| = |e \cap (V \setminus S)| = 1\}$$

(see Fig. 1.1), and the *size* of this cut is $|E(S, V \setminus S)|$, i.e., the number of edges. We also say that the cut is *induced* by S.

B. Gärtner and J. Matoušek, *Approximation Algorithms and Semidefinite Programming*, DOI 10.1007/978-3-642-22015-9_1,

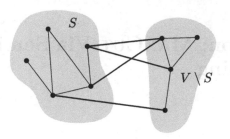

Fig. 1.1 The cut edges (*bold*) induced by a cut $(S, V \setminus S)$

The decision version of the MaxCut problem (given G and $k \in \mathbb{N}$, is there a cut of size at least k?) was shown to be NP-complete by Garey et al. [GJS76]. The above optimization version is consequently NP-hard.

1.2 Approximation Algorithms

Let us consider an optimization problem $\mathcal{P}$ (typically, but not necessarily, we will consider NP-hard problems). An *approximation algorithm* for $\mathcal{P}$ is a *polynomial-time algorithm* that computes a solution with some guaranteed quality for *every instance* of the problem. Here is a reasonably formal definition, formulated for maximization problems.

A maximization problem consists of a set $\mathcal{I}$ of *instances*. Every instance $I \in \mathcal{I}$ comes with a set $F(I)$ of *feasible solutions* (sometimes also called *admissible solutions*), and every $s \in F(I)$ in turn has a nonnegative real *value* $\omega(s) \geq 0$ associated with it. We also define

$$\mathrm{Opt}(I) = \sup_{s \in F(I)} \omega(s) \in \mathbb{R}_+ \cup \{-\infty, \infty\}$$

to be the optimum value of the instance. Value $-\infty$ occurs if $F(I) = \emptyset$, while $\mathrm{Opt}(I) = \infty$ means that there are feasible solutions of arbitrarily large value. To simplify the presentation, let us restrict our attention to problems where $\mathrm{Opt}(I)$ is finite for all I.

The MaxCut problem immediately fits into this setting. The instances are graphs, feasible solutions are subsets of vertices, and the value of a subset is the size of the cut induced by it.

1.2.1 Definition. *Let $\mathcal{P}$ be a maximization problem with set of instances $\mathcal{I}$, and let $\mathcal{A}$ be an algorithm that returns, for every instance $I \in \mathcal{I}$, a feasible solution $\mathcal{A}(I) \in F(I)$. Furthermore, let $\delta \colon \mathbb{N} \to \mathbb{R}_+$ be a function.*

We say that $\mathcal{A}$ is a δ-approximation algorithm for $\mathcal{P}$ if the following two conditions hold.

(i) *There exists a polynomial p such that for all $I \in \mathcal{I}$, the runtime of $\mathcal{A}$ on the instance I is bounded by $p(|I|)$, where $|I|$ is the encoding size of instance I.*

(ii) *For all instances $I \in \mathcal{I}$, $\omega(\mathcal{A}(I)) \geq \delta(|I|) \cdot \mathsf{Opt}(I)$.*

Encoding size is not a mathematically precise notion; what we mean is the following: For any given problem, we fix a reasonable "file format" in which we feed problem instances to the algorithm. For a graph problem such as MAXCUT, the format could be the number of vertices n, followed by a list of pairs of the form (i, j) with $1 \leq i < j \leq n$ that describe the edges. The encoding size of an instance can then be defined as the number of characters that are needed to write down the instance in the chosen format. Due to the fact that we allow runtime $p(|I|)$, where p is any polynomial, the precise format usually does not matter, and it is "reasonable" for every natural number k to be written down with $O(\log k)$ characters.

An interesting special case occurs when δ is a constant function. For $c \in \mathbb{R}$, a c-approximation algorithm is a δ-approximation algorithm with $\delta \equiv c$. Clearly, $c \leq 1$ must hold, and the closer c is to 1, the better the approximation.

We can smoothly extend the definition to randomized algorithms (algorithms that may use internal coin flips to guide their decisions). A randomized δ-approximation algorithm must have *expected* polynomial runtime and must satisfy

$$\mathbf{E}\left[\omega(\mathcal{A}(I))\right] \geq \delta(|I|) \cdot \mathsf{Opt}(I) \quad \text{for all } I \in \mathcal{I}.$$

For randomized algorithms , $\omega(\mathcal{A}(I))$ is a random variable, and we require that its *expectation* be a good approximation of the true optimum value.

For minimization problems, we replace sup by inf in the definition of $\mathsf{Opt}(I)$ and we require that $\omega(\mathcal{A}(I)) \leq \delta(|I|)\mathsf{Opt}(I)$ for all $I \in \mathcal{I}$. This leads to c-approximation algorithms with $c \geq 1$.

What Is Polynomial Time?

In the context of complexity theory, an algorithm is formally a Turing machine, and its runtime is obtained by counting the elementary operations (head movements), depending on the number of bits used to encode the problem on the input tape. This model of computation is also called the *bit model*.

The bit model is not very practical, and often the *real RAM* model, also called the *unit cost* model, is used instead.

The real RAM is a hypothetical computer, each of its memory cells capable of storing an arbitrary real number, including irrational ones like $\sqrt{2}$ or π.

Moreover, the model assumes that arithmetic operations on real numbers (including computations of square roots, trigonometric functions, random numbers, etc.) take constant time. The model is motivated by actual computers that approximate the real numbers by floating-point numbers with fixed precision.

The real RAM is a very convenient model, since it frees us from thinking about how to encode a real number, and what the resulting encoding size is. On the downside, the real RAM model is not always compatible with the Turing machine model. It can happen that we have a polynomial-time algorithm in the real RAM model, but when we translate it to a Turing machine, it becomes exponential.

For example, Gaussian elimination, one of the simplest algorithms in linear algebra, is not a polynomial-time algorithm in the Turing machine model if a naive implementation is used [GLS88, Sect. 1.4]. The reason is that in the naive implementation, intermediate results may require exponentially many bits.

Vice versa, a polynomial-time Turing machine may not be transferable to a polynomial-time real RAM algorithm. Indeed, the runtime of the Turing machine may tend to infinity with the encoding size of the input numbers, in which case there is no bound at all for the runtime that depends only on the *number* of input numbers.

In many cases, however, it *is* possible to implement a polynomial-time real RAM algorithm in such a way that all intermediate results have encoding lengths that are polynomial in the encoding lengths of the input numbers. In this case we also get a polynomial-time algorithm in the Turing machine model. For example, in the real RAM model, Gaussian elimination is an $O(n^3)$ algorithm for solving $n \times n$ linear equation systems. Using appropriate representations, it can be guaranteed that all intermediate results have bit lengths that are also polynomial in n [GLS88, Sect. 1.4], and we obtain that Gaussian elimination is a polynomial-time method also in the Turing machine model.

We will occasionally run into real RAM vs. Turing machine issues, and whenever we do so, we will try to be careful in sorting them out.

1.3 A Randomized 0.5-Approximation Algorithm for MaxCut

To illustrate previous definitions, let us describe a concrete (randomized) approximation algorithm `RandomizedMaxCut` for the MaxCut problem.

Given an instance $G = (V, E)$, the algorithm picks S as a random subset of V, where each vertex $v \in V$ is included in S with probability $1/2$, independent of all other vertices.

In a way this algorithm is stupid, since it never even looks at the edges. Still, we can prove the following result:

1.3.1 Theorem. *Algorithm* `RandomizedMaxCut` *is a randomized 0.5-approximation algorithm for the* MAXCUT *problem.*

Proof. It is clear that the algorithm runs in polynomial time. The value $\omega(\texttt{RandomizedMaxCut}(G))$ is the size of the cut (number of cut edges) generated by the algorithm (a random variable). Now we compute

$$
\begin{aligned}
\mathbf{E}\left[\omega(\texttt{RandomizedMaxCut}(G))\right] &= \mathbf{E}\left[|E(S, V \setminus S)|\right] \\
&= \sum_{e \in E} \mathrm{Prob}[e \in E(S, V \setminus S)] \\
&= \sum_{e \in E} \tfrac{1}{2} = \tfrac{1}{2}|E| \geq \tfrac{1}{2}\mathrm{Opt}(G).
\end{aligned}
$$

Indeed, $e \in E(S, V \setminus S)$ if and only if exactly one of the two endpoints of e ends up in S, and this has probability exactly $\frac{1}{2}$. $\qquad\square$

The main trick in this simple proof is to split the complicated-looking quantity $|E(S, V \setminus S)|$ into the contributions of individual edges, then we can use the linearity of expectation and account for the expected contribution of each edge separately. We will also see this trick in the analysis of the Goemans–Williamson algorithm.

It is possible to "derandomize" this algorithm and come up with a deterministic 0.5-approximation algorithm for MAXCUT (see Exercise 1.1). Minor improvements are possible. For example, there exists a $0.5(1 + 1/m)$ approximation algorithm, where $m = |E|$; see Exercise 1.2.

But until 1994, no c-approximation algorithm was found for any factor $c > 0.5$.

1.4 The Goemans–Williamson Algorithm

Here we describe the `GWMaxCut` algorithm, a 0.878-approximation algorithm for the MAXCUT problem, based on semidefinite programming. In a nutshell, a semidefinite program (SDP) is the problem of maximizing a linear function in n^2 variables x_{ij}, $i, j = 1, 2, \ldots, n$, subject to linear equality constraints *and* the requirement that the variables form a positive semidefinite matrix X. We write $X \succeq 0$ for "X is positive semidefinite."

For this chapter we assume that a semidefinite program can be solved in polynomial time, up to any desired accuracy ε, and under suitable conditions that are satisfied in our case. We refrain from specifying this further here; a detailed statement appears in Chap. 2. For now, let us continue with the

Goemans–Williamson approximation algorithm, using semidefinite programming as a black box.

We start by formulating the MAXCUT problem as a constrained optimization problem (which we will then turn into a semidefinite program). For the whole section, let us fix the graph $G = (V, E)$, where we assume that $V = \{1, 2, \ldots, n\}$ (this will be used often and in many places). Then we introduce variables $z_1, z_2, \ldots, z_n \in \{-1, 1\}$. Any assignment of values from $\{-1, 1\}$ to these variables encodes a cut $(S, V \setminus S)$, where $S = \{i \in V : z_i = 1\}$. The term

$$\frac{1 - z_i z_j}{2}$$

is exactly the contribution of the edge $\{i, j\}$ to the size of the above cut. Indeed, if $\{i, j\}$ is not a cut edge, we have $z_i z_j = 1$, and the contribution is 0. If $\{i, j\}$ is a cut edge, then $z_i z_j = -1$, and the contribution is 1. It follows that we can reformulate the MAXCUT problem as follows.

$$\text{Maximize} \quad \sum_{\{i,j\} \in E} \frac{1 - z_i z_j}{2} \qquad \qquad (1.1)$$
$$\text{subject to} \quad z_i \in \{-1, 1\}, \quad i = 1, \ldots, n.$$

The optimum value (or simply value) of this program is $\mathsf{Opt}(G)$, the size of a maximum cut. Thus, in view of the NP-completeness of MAXCUT, we cannot expect to solve this optimization problem exactly in polynomial time.

Semidefinite Programming Relaxation

Here is the crucial step: We write down a semidefinite program whose value is an *upper bound* for the value $\mathsf{Opt}(G)$ of (1.1). To get it, we first replace each real variable z_i with a vector variable $\mathbf{u}_i \in S^{n-1} = \{\mathbf{x} \in \mathbb{R}^n : \|\mathbf{x}\| = 1\}$, the $(n - 1)$-dimensional unit sphere:

$$\text{Maximize} \quad \sum_{\{i,j\} \in E} \frac{1 - \mathbf{u}_i^T \mathbf{u}_j}{2} \qquad \qquad (1.2)$$
$$\text{subject to} \quad \mathbf{u}_i \in S^{n-1}, \quad i = 1, 2, \ldots, n.$$

This is called a *vector program* since the unknowns are vectors.[1]

From the fact that the set $\{-1, 1\}$ can be embedded into S^{n-1} via the mapping $x \mapsto (0, 0, \ldots, 0, x)$, we derive the following important property: for every solution of (1.1), there is a corresponding solution of (1.2) with the same value. This means that the program (1.2) is a *relaxation* of (1.1), a program with "more" solutions, and it therefore has value *at least* $\mathsf{Opt}(G)$. It is also

---

[1] We consider vectors in $\mathbb{R}^n$ as column vectors, i.e., as $n \times 1$ matrices. The superscript ^{T} denotes matrix transposition, and thus $\mathbf{u}_i^T \mathbf{u}_j$ is the standard scalar product of $\mathbf{u}_i$ and $\mathbf{u}_j$.

clear that this value is still finite, since $\mathbf{u}_i^T \mathbf{u}_j$ is bounded from below by -1 for all i, j.

Vectors may look more complicated than real numbers, and so it is quite counterintuitive that (1.2) should be any easier than (1.1). But semidefinite programming will allow us to solve the vector program efficiently, to any desired accuracy!

To see this, we perform yet another variable substitution, namely, $x_{ij} = \mathbf{u}_i^T \mathbf{u}_j$. This brings (1.2) into the form of a semidefinite program:

$$
\begin{aligned}
\text{Maximize} \quad & \sum_{\{i,j\} \in E} \frac{1 - x_{ij}}{2} \\
\text{subject to} \quad & x_{ii} = 1, \quad i = 1, 2, \ldots, n, \\
& X \succeq 0.
\end{aligned} \tag{1.3}
$$

To see that (1.3) is equivalent to (1.2), we first note that if $\mathbf{u}_1, \ldots, \mathbf{u}_n$ constitute a feasible solution to (1.2), i.e., they are unit vectors, then with $x_{ij} = \mathbf{u}_i^T \mathbf{u}_j$, we have

$$X = U^T U,$$

where the matrix U has the columns $\mathbf{u}_1, \mathbf{u}_2, \ldots, \mathbf{u}_n$. Such a matrix X is positive semidefinite, and $x_{ii} = 1$ follows from $\mathbf{u}_i \in S^{n-1}$ for all i. So X is a feasible solution of (1.3) with the same value.

Slightly more interesting is the opposite direction, namely, that every feasible solution X of (1.3) yields a solution of (1.2), with the same value. For this, one needs to know that every positive semidefinite matrix X can be written as the product $X = U^T U$ (see Sect. 2.2). Thus, if X is a feasible solution of (1.3), the columns of such a matrix U provide a feasible solution of (1.2); due to the constraints $x_{ii} = 1$, they are actually unit vectors.

Thus, the semidefinite program (1.3) has the same finite value $\mathsf{SDP}(G) \geq \mathsf{Opt}(G)$ as (1.2). So we can find in polynomial time a matrix $X^* \succeq 0$ with $x_{ii}^* = 1$ for all i and with

$$\sum_{\{i,j\} \in E} \frac{1 - x_{ij}^*}{2} \geq \mathsf{SDP}(G) - \varepsilon,$$

for every $\varepsilon > 0$.

We can also compute in polynomial time a matrix U^* such that $X^* = (U^*)^T U^*$, up to a tiny error. This is a *Cholesky factorization* of X^*; see Sect. 2.3. The tiny error can be dealt with at the cost of slightly adapting ε. So let us assume that the factorization is exact.

Then the columns $\mathbf{u}_1^*, \mathbf{u}_2^*, \ldots, \mathbf{u}_n^*$ of U^* are unit vectors that form an almost-optimal solution of the vector program (1.2):

$$\sum_{\{i,j\} \in E} \frac{1 - \mathbf{u}_i^{*T} \mathbf{u}_j^*}{2} \geq \mathsf{SDP}(G) - \varepsilon \geq \mathsf{Opt}(G) - \varepsilon. \tag{1.4}$$

Rounding the Vector Solution

Let us recall that what we actually want to solve is program (1.1), where the n variables z_i are elements of $S^0 = \{-1,1\}$ and thus determine a cut $(S, V \setminus S)$ via $S := \{i \in V : z_i = 1\}$.

What we have is an almost optimal solution of the relaxed program (1.2) where the n vector variables are elements of S^{n-1}. We therefore need a way of mapping S^{n-1} back to S^0 in such a way that we do not "lose too much." Here is how we do it. Choose $\mathbf{p} \in S^{n-1}$ and consider the mapping

$$\mathbf{u} \mapsto \begin{cases} 1 & \text{if } \mathbf{p}^T\mathbf{u} \geq 0, \\ -1 & \text{otherwise.} \end{cases} \tag{1.5}$$

The geometric picture is the following: $\mathbf{p}$ partitions S^{n-1} into a closed hemisphere $H = \{\mathbf{u} \in S^{n-1} : \mathbf{p}^T\mathbf{u} \geq 0\}$ and its complement. Vectors in H are mapped to 1, while vectors in the complement map to -1; see Fig. 1.2.

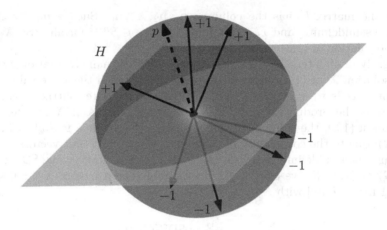

Fig. 1.2 Rounding vectors in S^{n-1} to $\{-1, 1\}$ through a vector $\mathbf{p} \in S^{n-1}$

It remains to choose $\mathbf{p}$, and we will do this *randomly* (we speak of *randomized rounding*). More precisely, we sample $\mathbf{p}$ uniformly at random from S^{n-1}. To understand why this is a good thing, we need to do the computations, but here is the intuition. We certainly want that a pair of vectors $\mathbf{u}_i^*$ and $\mathbf{u}_j^*$ with large value

$$\frac{1 - \mathbf{u}_i^{*T}\mathbf{u}_j^*}{2}$$

is more likely to yield a cut edge $\{i, j\}$ than a pair with a small value. Since the contribution grows with the angle between $\mathbf{u}_i^*$ and $\mathbf{u}_j^*$, our mapping to

$\{-1, +1\}$ should be such that pairs with large angles are more likely to be mapped to different values than pairs with small angles.

As we will see, this is how the function (1.5) with randomly chosen $\mathbf{p}$ is going to behave.

1.4.1 Lemma. *Let* $\mathbf{u}, \mathbf{u}' \in S^{n-1}$. *The probability that (1.5) maps* $\mathbf{u}$ *and* $\mathbf{u}'$ *to different values is*

$$\frac{1}{\pi} \arccos \mathbf{u}^T \mathbf{u}'.$$

Proof. Let $\alpha \in [0, \pi]$ be the angle between the unit vectors $\mathbf{u}$ and $\mathbf{u}'$. By the law of cosines, we have

$$\cos(\alpha) = \mathbf{u}^T \mathbf{u}' \in [-1, 1],$$

or, in other words,

$$\alpha = \arccos \mathbf{u}^T \mathbf{u}' \in [0, \pi].$$

If $\alpha = 0$ or $\alpha = \pi$, meaning that $\mathbf{u} \in \{\mathbf{u}', -\mathbf{u}'\}$, the statement trivially holds. Otherwise, let us consider the linear span L of $\mathbf{u}$ and $\mathbf{u}'$, which is a two-dimensional subspace of $\mathbb{R}^n$. With $\mathbf{r}$ the projection of $\mathbf{p}$ to that subspace, we have $\mathbf{p}^T \mathbf{u} = \mathbf{r}^T \mathbf{u}$ and $\mathbf{p}^T \mathbf{u}' = \mathbf{r}^T \mathbf{u}'$. This means that $\mathbf{u}$ and $\mathbf{u}'$ map to different values if and only if $\mathbf{r}$ lies in a "half-open double wedge" W of opening angle α; see Fig. 1.3.

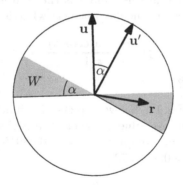

Fig. 1.3 Randomly rounding vectors: $\mathbf{u}$ and $\mathbf{u}'$ map to different values if and only if the projection $\mathbf{r}$ of $\mathbf{p}$ to the linear span of $\mathbf{u}$ and $\mathbf{u}'$ lies in the shaded region W ("half-open double wedge")

Since $\mathbf{p}$ is uniformly distributed in S^{n-1}, the direction of $\mathbf{r}$ is uniformly distributed in $[0, 2\pi]$. Therefore, the probability of $\mathbf{r}$ falling into the double wedge is the fraction of angles covered by the double wedge, and this is α/π. $\qquad\square$

Getting the Bound

Let us see what we have achieved. If we round as above, the expected number of edges in the resulting cut equals

$$\sum_{\{i,j\}\in E} \frac{\arccos \mathbf{u}_i^{*T} \mathbf{u}_j^*}{\pi}.$$

Indeed, we are summing the probability that an edge $\{i,j\}$ becomes a cut edge, as in Lemma 1.4.1, over all edges $\{i,j\}$. The trouble is that we do not know much about this sum. But we *do* know that

$$\sum_{\{i,j\}\in E} \frac{1 - \mathbf{u}_i^{*T} \mathbf{u}_j^*}{2} \geq \mathsf{Opt}(G) - \varepsilon;$$

see (1.4). The following technical lemma allows us to compare the two sums termwise.

1.4.2 Lemma. *For all $z \in [-1,1]$,*

$$\frac{\arccos(z)}{\pi} \geq 0.8785672 \, \frac{1-z}{2}.$$

The constant appearing in this lemma is the solution to a problem that seems to come from a crazy calculus teacher: what is the minimum of the function

$$f(z) = \frac{2\arccos(z)}{\pi(1-z)}$$

over the interval $[-1,1]$?

Proof. The plot in Fig. 1.4 below depicts the function $f(z)$; the minimum occurs at the (unique) value z^* where the derivative vanishes. Using a numeric solver, you can compute $z^* \approx -0.68915773665$, which yields $f(z^*) \approx 0.87856720578 > 0.8785672$. □

Using this lemma, we can conclude that the expected number of cut edges produced by our algorithm satisfies

$$\sum_{\{i,j\}\in E} \frac{\arccos \mathbf{u}_i^{*T} \mathbf{u}_j^*}{\pi} \geq 0.8785672 \sum_{\{i,j\}\in E} \frac{1 - \mathbf{u}_i^{*T} \mathbf{u}_j^*}{2}$$
$$\geq 0.8785672(\mathsf{Opt}(G) - \varepsilon)$$
$$\geq 0.878\mathsf{Opt}(G),$$

provided we choose $\varepsilon \leq 5 \cdot 10^{-4}$.

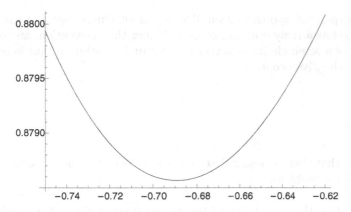

Fig. 1.4 The function $f(z) = 2\arccos(z)/\pi(1-z)$ and its minimum

Here is a summary of the Goemans–Williamson algorithm `GWMaxCut` for approximating the maximum cut in a graph $G = (\{1, 2, \ldots, n\}, E)$.

1. Compute an almost optimal solution $\mathbf{u}_1^*, \mathbf{u}_2^*, \ldots, \mathbf{u}_n^*$ of the vector program

$$\text{maximize} \quad \sum_{\{i,j\}\in E} \frac{1-\mathbf{u}_i^T\mathbf{u}_j}{2}$$
$$\text{subject to} \quad \mathbf{u}_i \in S^{n-1}, \quad i = 1, 2, \ldots, n.$$

This is a solution that satisfies

$$\sum_{\{i,j\}\in E} \frac{1 - \mathbf{u}_i^{*T}\mathbf{u}_j^*}{2} \geq \mathsf{SDP}(G) - 5\cdot 10^{-4} \geq \mathsf{Opt}(G) - 5\cdot 10^{-4},$$

and it can be found in polynomial time by semidefinite programming and Cholesky factorization.

2. Choose $\mathbf{p} \in S^{n-1}$ uniformly at random, and output the cut induced by

$$S := \{i \in \{1, 2, \ldots, n\}: \mathbf{p}^T\mathbf{u}_i^* \geq 0\}.$$

We have thus proved the following result.

1.4.3 Theorem. *Algorithm* `GWMaxCut` *is a randomized* 0.878-*approximation algorithm for the* MAXCUT *problem.*

Almost optimal vs. optimal solutions. It is customary in the literature (and we will adopt this later) to simply call an almost optimal solution of a semidefinite or a vector program an "optimal solution." This is justified, since for the purpose of approximation algorithms an almost optimal solution is just as good as a truly optimal solution. Under this convention, an "optimal solution" of a semidefinite or a vector program is a solution that is accurate enough in the given context.

Exercises

1.1 Prove that there is also a deterministic 0.5-approximation algorithm for the MAXCUT problem.

1.2 Prove that there is a $0.5(1+1/m)$-approximation algorithm (randomized or deterministic) for the MAXCUT problem, where m is the number of edges of the given graph G.

Chapter 2
Semidefinite Programming

Let us start with the concept of *linear programming*. A *linear program* is the problem of maximizing (or minimizing) a linear function in n variables subject to linear equality and inequality constraints. In *equational form*, a linear program can be written as

$$\text{maximize} \quad \mathbf{c}^T \mathbf{x}$$
$$\text{subject to} \quad A\mathbf{x} = \mathbf{b}$$
$$\mathbf{x} \geq \mathbf{0}.$$

Here $\mathbf{x} = (x_1, x_2, \ldots, x_n)$ is a vector of n variables,[1] $\mathbf{c} = (c_1, c_2, \ldots, c_n)$ is the objective function vector, $\mathbf{b} = (b_1, b_2, \ldots, b_m)$ is the right-hand side, and $A \in \mathbb{R}^{m \times n}$ is the constraint matrix. The bold digit $\mathbf{0}$ stands for the zero vector of the appropriate dimension. Vector inequalities like $\mathbf{x} \geq \mathbf{0}$ are to be understood componentwise.

In other words, among all $\mathbf{x} \in \mathbb{R}^n$ that satisfy the matrix equation $A\mathbf{x} = \mathbf{b}$ and the vector inequality $\mathbf{x} \geq \mathbf{0}$ (such $\mathbf{x}$ are called *feasible solutions*), we are looking for an $\mathbf{x}^*$ with the highest value $\mathbf{c}^T \mathbf{x}^*$.

2.1 From Linear to Semidefinite Programming

To get a semidefinite program, we replace the vector space $\mathbb{R}^n$ underlying $\mathbf{x}$ by another real vector space, namely the vector space

$$\mathrm{SYM}_n = \{ X \in \mathbb{R}^{n \times n} : x_{ij} = x_{ji}, 1 \leq i < j \leq n \}$$

of symmetric $n \times n$ matrices, and we replace the matrix A by a linear mapping $A \colon \mathrm{SYM}_n \to \mathbb{R}^m$.

---

[1] Vectors are column vectors, but in writing them explicitly, we use the n-tuple notation.

B. Gärtner and J. Matoušek, *Approximation Algorithms and Semidefinite Programming*, DOI 10.1007/978-3-642-22015-9_2,
© Springer-Verlag Berlin Heidelberg 2012

The standard scalar product $\langle \mathbf{x}, \mathbf{y} \rangle = \mathbf{x}^T \mathbf{y}$ over $\mathbb{R}^n$ gets replaced by the standard scalar product

$$X \bullet Y := \sum_{i=1}^{n} \sum_{j=1}^{n} x_{ij} y_{ij}$$

over SYM_n. Alternatively, we can also write $X \bullet Y = \mathrm{Tr}(X^T Y)$, where for a square matrix M, $\mathrm{Tr}(M)$ (the *trace* of M) is the sum of the diagonal entries of M.

Finally, we replace the constraint $\mathbf{x} \geq \mathbf{0}$ by the constraint

$$X \succeq 0.$$

Here $X \succeq 0$ stands for "the matrix X is positive semidefinite."

Next, we will explain all of this in more detail.

2.2 Positive Semidefinite Matrices

First we recall that a *positive semidefinite matrix* is a real matrix M that is *symmetric* (i.e., $M^T = M$, and in particular, M is a square matrix) and has all eigenvalues *nonnegative*. (The condition of symmetry is all too easy to forget. Let us also recall from Linear Algebra that a symmetric real matrix has only real eigenvalues, and so the nonnegativity condition makes sense.)

Here are several equivalent characterizations.

2.2.1 Fact. *Let $M \in \mathrm{SYM}_n$. The following statements are equivalent.*

(i)   *M is positive semidefinite, i.e., all the eigenvalues of M are nonnegative.*
(ii)   *$\mathbf{x}^T M \mathbf{x} \geq 0$ for all $\mathbf{x} \in \mathbb{R}^n$.*
(iii) *There exists a matrix $U \in \mathbb{R}^{n \times n}$ such that $M = U^T U$.*

This can easily be proved using diagonalization, which is a basic tool for dealing with symmetric matrices.

Using the condition (ii), we can see that a semidefinite program as introduced earlier can be regarded as a "linear program with infinitely many constraints." Indeed, the constraint $X \succeq 0$ for the unknown matrix X can be replaced with the constraints $\mathbf{a}^T X \mathbf{a} \geq 0$, $\mathbf{a} \in \mathbb{R}^n$. That is, we have infinitely many linear constraints, one for every vector $\mathbf{a} \in \mathbb{R}^n$.

2.2.2 Definition. PSD_n *is the set of all positive semidefinite $n \times n$ matrices.*

A matrix M is called *positive definite* if $\mathbf{x}^T M \mathbf{x} > 0$ for all $\mathbf{x} \neq \mathbf{0}$. It can be checked that the positive definite matrices form the interior of the set $\text{PSD}_n \subseteq \text{SYM}_n$.

2.3 Cholesky Factorization

In semidefinite programming we often need to compute, for a given positive semidefinite matrix M, a matrix U as in Fact 2.2.1(iii), i.e., such that $M = U^T U$. This is called the computation of a *Cholesky factorization*. (The definition also requires U to be upper triangular, but we don't need this.)

We present a simple explicit method, the *outer product Cholesky Factorization* [GvL96, Sect. 4.2.8], which uses $O(n^3)$ arithmetic operations for an $n \times n$ matrix M.

If $M = (\alpha) \in \mathbb{R}^{1 \times 1}$, we set $U = (\sqrt{\alpha})$, where $\alpha \geq 0$ by the nonnegativity of the eigenvalues. Otherwise, since M is symmetric, we can write it as

$$M = \begin{pmatrix} \alpha & \mathbf{q}^T \\ \mathbf{q} & N \end{pmatrix}.$$

We also have $\alpha = \mathbf{e}_1^T M \mathbf{e}_1 \geq 0$ by Fact 2.2.1(ii). Here $\mathbf{e}_i$ denotes the i-th unit vector of the appropriate dimension.

There are two cases to consider. If $\alpha > 0$, we compute

$$M = \begin{pmatrix} \sqrt{\alpha} & \mathbf{0}^T \\ \frac{1}{\sqrt{\alpha}} \mathbf{q} & I_{n-1} \end{pmatrix} \begin{pmatrix} 1 & \mathbf{0}^T \\ \mathbf{0} & N - \frac{1}{\alpha} \mathbf{q} \mathbf{q}^T \end{pmatrix} \begin{pmatrix} \sqrt{\alpha} & \frac{1}{\sqrt{\alpha}} \mathbf{q}^T \\ \mathbf{0} & I_{n-1} \end{pmatrix}. \qquad (2.1)$$

The matrix $N - \frac{1}{\alpha} \mathbf{q} \mathbf{q}^T$ is again positive semidefinite (Exercise 2.2), and we can recursively compute a Cholesky factorization

$$N - \frac{1}{\alpha} \mathbf{q} \mathbf{q}^T = V^T V.$$

Elementary calculations yield that

$$U = \begin{pmatrix} \sqrt{\alpha} & \frac{1}{\sqrt{\alpha}} \mathbf{q}^T \\ \mathbf{0} & V \end{pmatrix}$$

satisfies $M = U^T U$, and so we have found a Cholesky factorization of M.

In the other case ($\alpha = 0$), we also have $\mathbf{q} = \mathbf{0}$ (Exercise 2.2). The matrix N is positive semidefinite (apply Fact 2.2.1(ii) with $\mathbf{x} = (0, x_2, \ldots, x_n)$), so we can recursively compute a matrix V satisfying $N = V^T V$. Setting

$$U = \begin{pmatrix} 0 & \mathbf{0}^T \\ \mathbf{0} & V \end{pmatrix}$$

then gives $M = U^T U$, and we are done with the outer product Cholesky factorization.

Exercise 2.3 asks you to show that the above method can be modified to check whether a given matrix M is positive semidefinite.

We note that the outer product Cholesky factorization is a polynomial-time algorithm only in the real RAM model. We can transform it into a polynomial-time Turing machine, but at the cost of giving up the exact factorization. After all, a Turing machine cannot even exactly factor the 1×1 matrix (2), since $\sqrt{2}$ is an irrational number that cannot be written down with finitely many bits.

The error analysis of Higham [Hig91] implies the following: when we run a modified version of the above algorithm (the modification is to base the factorization (2.1) not on $\alpha = m_{11}$ but rather on the largest diagonal entry m_{jj}), and when we round all intermediate results to $O(n)$ bits (the constant chosen appropriately), then we will obtain a matrix U such that the relative error $\|U^T U - M\|_F / \|M\|_F$ is bounded by 2^{-n}. (Here $\|M\|_F = \left(\sum_{i,j=1}^n m_{ij}^2\right)^{1/2}$ is the *Frobenius norm*.) This accuracy is sufficient for most purposes, and in particular, for the Goemans–Williamson MAXCUT algorithm of the previous chapter.

2.4 Semidefinite Programs

2.4.1 Definition. *A semidefinite program in equational form is the following kind of optimization problem:*

$$
\begin{aligned}
\text{Maximize} \quad & \sum_{i,j=1}^n c_{ij} x_{ij} \\
\text{subject to} \quad & \sum_{i,j=1}^n a_{ijk} x_{ij} = b_k, \quad k = 1, \dots, m, \\
& X \succeq 0,
\end{aligned}
\tag{2.2}
$$

where the x_{ij}, $1 \le i, j \le n$, are n^2 variables satisfying the symmetry conditions $x_{ji} = x_{ij}$ for all i, j, the c_{ij}, a_{ijk} and b_k are real coefficients, and

$$
X = (x_{ij})_{i,j=1}^n \in \mathrm{SYM}_n.
$$

In a more compact form, the semidefinite program in this definition can be written as

$$
\begin{aligned}
\text{Maximize} \quad & C \bullet X \\
\text{subject to} \quad & A_1 \bullet X = b_1 \\
& A_2 \bullet X = b_2 \\
& \qquad \vdots \\
& A_m \bullet X = b_m \\
& X \succeq 0,
\end{aligned}
\tag{2.3}
$$

where
$$C = (c_{ij})_{i,j=1}^{n}$$
is the matrix expressing the objective function,[2] and

$$A_k = (a_{ijk})_{i,j=1}^{n}, \quad k = 1, 2, \ldots, m.$$

(We recall the notation $C \bullet X = \sum_{i,j=1}^{n} c_{ij}x_{ij}$ introduced earlier.)

We can write the system of m linear constraints $A_1 \bullet X = b_1, \ldots, A_m \bullet X = b_m$ even more compactly as
$$A(X) = \mathbf{b},$$
where $\mathbf{b} = (b_1, \ldots, b_m)$ and $A \colon \mathrm{SYM}_n \mapsto \mathbb{R}^m$ is a linear mapping. This notation will be useful especially for general considerations about semidefinite programs.

Following the linear programming case, we call the semidefinite program (2.3) *feasible* if there is some *feasible solution*, i.e., a matrix $\tilde{X} \in \mathrm{SYM}_n$ with $A(\tilde{X}) = \mathbf{b}$, $\tilde{X} \succeq 0$. The *value* of a feasible semidefinite program is defined as

$$\sup\{C \bullet X : A(X) = \mathbf{b}, X \succeq 0\}, \tag{2.4}$$

which includes the possibility that the value is ∞. In this case, the program is called *unbounded*, otherwise, we speak of a *bounded* semidefinite program.

An *optimal solution* is a feasible solution X^* such that $C \bullet X^* \geq C \bullet X$ for all feasible solutions X. Consequently, if there is an optimal solution, the value of the semidefinite program is finite, and it is attained, meaning that the supremum in (2.4) is a maximum.

Warning: If a semidefinite program has finite value, generally we cannot conclude that the value is attained! We illustrates this with an example below. For applications, this presents no problem: All known efficient algorithms for solving semidefinite programs return only *approximately optimal* solutions, and these are the ones that we rely on in applications.

Here is the example. With $X \in \mathrm{SYM}_2$, let us consider the problem

$$
\begin{aligned}
\text{Maximize} \quad & -x_{11} \\
\text{subject to} \quad & x_{12} = 1 \\
& X \succeq 0.
\end{aligned}
$$

The feasible solutions of this semidefinite program are all positive semidefinite matrices X of the form
$$X = \begin{pmatrix} x_{11} & 1 \\ 1 & x_{22} \end{pmatrix}.$$

---

[2] Since X is symmetric, we may also assume that C is symmetric, without loss of generality; similarly for the matrices A_k.

It is easy to see that such a matrix is positive semidefinite if and only if $x_{11}, x_{22} \geq 0$ and $x_{11} x_{22} \geq 1$. Equivalently, if $x_{11} > 0$ and $x_{22} \geq 1/x_{11}$. This implies that the value of the program is 0, but there is no solution that attains this value.

2.5 Non-standard Form

Semidefinite programs do not always look exactly as in (2.3). Besides the constraints given by linear equations, as in (2.3), there may also be inequality constraints, and one may also need extra real variables that are not entries of the positive semidefinite matrix X. Let us indicate how such more general semidefinite programs can be converted to the standard form (2.3).

First, introducing extra nonnegative real variables $x_1, x_2, \ldots, x_k$ not appearing in X can be handled by incorporating them into the matrix. Namely, we replace X with the matrix $X' \in \mathrm{SYM}_{n+k}$, of the form

$$
X' = \begin{pmatrix}
X & 0 & 0 & \cdots & 0 \\
0 & x_1 & 0 & \cdots & 0 \\
0 & 0 & x_2 & \cdots & 0 \\
0 & 0 & 0 & \ddots & 0 \\
0 & 0 & 0 & \cdots & x_k
\end{pmatrix}.
$$

We note that the zero entries really mean adding equality constraints to the standard form (2.3). We have $X' \succeq 0$ if and only if $X \succeq 0$ and $x_1, x_2, \ldots, x_k \geq 0$.

To get rid of inequalities, we can add nonnegative slack variables, just as in linear programming. Thus, an inequality constraint $x_{23} + 5x_{15} \leq 22$ is replaced with the equality constraint $x_{23} + 5x_{15} + y = 22$, where y is an extra nonnegative real variable that does not occur anywhere else. Finally, an unrestricted real variable x_i (allowed to attain both positive and negative values) is replaced by the difference $x_i' - x_i''$, where x_i' and x_i'' are two new *nonnegative* real variables.

By these steps, a non-standard semidefinite program assumes the form of a standard program (2.3) over SYM_{n+k} for some k.

2.6 The Complexity of Solving Semidefinite Programs

In Chap. 1 we claimed that under suitable conditions, satisfied in the Goemans–Williamson MaxCut algorithm and many other applications, a semidefinite program can be solved in polynomial time up to any desired accuracy ε. Here we want to make this claim precise.

In order to claim that a semidefinite program is (approximately) solvable in polynomial time, we need to assume that it is "well-behaved" in some sense. Namely, we need that the feasible solutions cannot be too large: we will assume that together with the input semidefinite program, we also obtain an integer R bounding the Frobenius norm of all feasible matrices X.

We will be able to claim polynomial-time approximate solvability only in the case where R has polynomially many digits. As we will see later, one can construct examples of semidefinite programs where this fails and one needs exponentially many bits in order to write down any feasible solution.

What the ellipsoid method can do. The strongest known *theoretical* result on solvability of semidefinite programs follows from the *ellipsoid method* (a standard reference is Grötschel et al. [GLS88]). The ellipsoid method is a general algorithm for maximizing (or minimizing) a given linear function over a given *full-dimensional* convex set C.[3]

In our case, we would like to apply the ellipsoid method to the set $C \subseteq$ SYM_n of all feasible solutions of the considered semidefinite program.

This set C is convex but not full-dimensional, due to the linear equality constraints in the semidefinite program. But since the affine solution space L of the set of linear equalities can be computed in polynomial time through Gaussian elimination, we may restrict C to this space and then we have a full-dimensional convex set. Technically, this can either be done through an explicit coordinate transformation, or dealt with implicitly (we will do the latter).

The ellipsoid method further requires that C should be enclosed in a ball of radius R and it should be given by a polynomial-time *weak separation oracle* [GLS88, Sect. 2.1]. In our case, this means that for a given symmetric matrix X that satisfies all the equality constraints, we can either certify that it is "almost" feasible (i.e., has small distance to the set PSD_n), or find a hyperplane that almost separates X from C. Polynomial time is w.r.t. the encoding length of X, the bound R, and the amount of "almost."

It turns out that a polynomial-time weak separation oracle is provided by the Cholesky factorization algorithm (see Sect. 2.3 and Exercise 2.3). The only twist is that we need to perform the decomposition "within" L, i.e., for a suitably transformed matrix X' of lower dimension.

Indeed, if the approximate Cholesky factorization goes through, X' is an almost positive semidefinite matrix, since it is close (in absolute terms) to a positive semidefinite matrix $U^T U$. The outer product Cholesky factorization guarantees a small *relative* error, but this can be turned into a small absolute error by computing with $O(\log R)$ more bits.

Similarly, if the approximate Cholesky factorization fails at some point, we can reconstruct a vector $\mathbf{v}$ (by solving a system of linear equations) such that $\mathbf{v}^T X' \mathbf{v}$ is negative or at least very close to zero; this gives us an almost separating hyperplane.

---

[3] A set C is convex if for all $\mathbf{x}, \mathbf{y} \in C$ and $\lambda \in [0, 1]$, we also have $(1 - \lambda)\mathbf{x} + \lambda \mathbf{y} \in C$.

To state the result, we consider a semidefinite program (P) in the form

$$\begin{aligned}
\text{Maximize} \quad & C \bullet X \\
\text{subject to} \quad & A_1 \bullet X = b_1 \\
& A_2 \bullet X = b_2 \\
& \vdots \\
& A_m \bullet X = b_m \\
& X \succeq 0.
\end{aligned}$$

Let $L := \{X \in \mathrm{SYM}_n : A_i \bullet X = b_i, i = 1, 2, \ldots, m\}$ be the affine subspace of matrices satisfying all the equality constraints. Let us say that a matrix $X \in \mathrm{SYM}_n$ is an ε-*deep feasible solution* of (P) if all matrices $Y \in L$ of (Frobenius) distance at most ε from X are feasible solutions of (P).

Now we can state a precise result about the solvability of semidefinite programs, which follows from general results about the ellipsoid method [GLS88, Theorem 3.2.1. and Corollary 4.2.7].

2.6.1 Theorem. *Let us assume that the semidefinite program (P) has rational coefficients, let R be an explicitly given bound on the maximum Frobenius norm $\|X\|_F$ of all feasible solutions of (P), and let $\varepsilon > 0$ be a rational number.*

Let us put $v_{\mathrm{deep}} := \sup\{C \bullet X : X \text{ an } \varepsilon\text{-deep feasible solution of } (P)\}$. There is an algorithm, with runtime polynomial in the (binary) encoding sizes of the input numbers and in $\log(R/\varepsilon)$, that produces one of the following two outputs.

(a) *A matrix $X^* \in L$ (i.e., satisfying all equality constraints) such that $\|X^* - X\|_F \le \varepsilon$ for some feasible solution X, and with $C \bullet X^* \ge v_{\mathrm{deep}} - \varepsilon$.*

(b) *A certificate that (P) has no ε-deep feasible solutions. This certificate has the form of an ellipsoid $E \subset L$ that, on the one hand, is guaranteed to contain all feasible solutions, and on the other hand, has volume so small that it cannot contain an ε-ball.*

One has to be careful here: This theorem does not yet imply the informal claim made in Chap. 1. It does so if R is not too large. Unfortunately, R may have to be very large in general, namely doubly-exponential in n, the matrix size; see the pathological example below. In such a case, the bound of Theorem 2.6.1 is exponential!

What saves us in the applications is that R is usually small. In the MAX-CUT application, for example, all entries of a feasible solution X are inner products of unit vectors. Hence the entries are in $[-1, 1]$, and thus $\|X\|_F \le n$.

A glance at other algorithms. First we want to point out that the ellipsoid method is the *only known* method that provably yields polynomial runtime in the Turing machine model, at least under suitable and fairly general conditions such as a good bound R.

On the other hand, the practical performance of the ellipsoid method is poor, and completely different algorithms have made semidefinite programming into an extremely powerful computational tool in practice.

Perhaps the most significant and most widely used class of algorithms are *interior-point methods*, which we will outline in Chap. 6. On the theoretical side, they are capable of providing polynomial-time bounds in the RAM model, but there is no control over the sizes of the intermediate numbers that come up in the computations, as far as we could find in the (huge) literature. Moreover, describing these methods in full detail is beyond the scope of this book.

In order to provide a simple and complete algorithm for semidefinite programming, we will present and analyze *Hazan's algorithm* in Chap. 5. This is a recent alternative method for approximately solving semidefinite programs, with a polynomial bound on the running time in the real RAM model. It comes with output guarantees similar to the ones in Theorem 2.6.1 above, and it is efficient in practice. However, the running time bound is polynomial only in $1/\varepsilon$ and not in $\log(1/\varepsilon)$.

A semidefinite program where all feasible solutions are huge. To get such a pathological example, let us consider a semidefinite program with the following constraints:

$$
\begin{pmatrix}
1 & 2 & 0 & 0 & 0 & 0 & \cdots & 0 & 0 \\
2 & x_1 & 0 & 0 & 0 & 0 & \cdots & 0 & 0 \\
0 & 0 & 1 & x_1 & 0 & 0 & \cdots & 0 & 0 \\
0 & 0 & x_1 & x_2 & 0 & 0 & \cdots & 0 & 0 \\
0 & 0 & 0 & 0 & 1 & x_2 & \cdots & 0 & 0 \\
0 & 0 & 0 & 0 & x_2 & x_3 & \cdots & 0 & 0 \\
& & & \vdots & & & \ddots & & \vdots \\
0 & 0 & 0 & 0 & 0 & 0 & \cdots & 1 & x_{n-1} \\
0 & 0 & 0 & 0 & 0 & 0 & \cdots & x_{n-1} & x_n
\end{pmatrix} \succeq 0.
$$

This is in fact a constraint of the form $X \succeq 0$, along with various equalities involving entries of X. Due to the block structure, we have $X \succeq 0$ if and only if

$$
\begin{pmatrix} 1 & x_{i-1} \\ x_{i-1} & x_i \end{pmatrix} \succeq 0, \quad i = 1, \ldots, n,
$$

where $x_0 := 2$. But this implies

$$
\det \begin{pmatrix} 1 & x_{i-1} \\ x_{i-1} & x_i \end{pmatrix} = x_i - x_{i-1}^2 \geq 0, \quad i = 1, \ldots, n,
$$

equivalently $x_i \geq x_{i-1}^2$, $i = 1, \ldots, n$. It follows that

$$x_n \geq 2^{2^n}$$

for every feasible solution, which is doubly-exponential in n. Hence, the encoding size of x_n (when written as say a rational number) is exponential in n and also in the number of variables.

Exercises

2.1 Prove or disprove the following claim: For all $A, B \in \mathrm{SYM}_n$, we also have $AB \in \mathrm{SYM}_n$.

2.2 Fill in the missing details of the outer product Cholesky factorization.

(i) If the matrix

$$M = \begin{pmatrix} \alpha & \mathbf{q}^T \\ \mathbf{q} & N \end{pmatrix}$$

is positive semidefinite with $\alpha > 0$, then the matrix

$$N - \frac{1}{\alpha} \mathbf{q} \mathbf{q}^T$$

is also positive semidefinite.

(ii) If the matrix

$$M = \begin{pmatrix} 0 & \mathbf{q}^T \\ \mathbf{q} & N \end{pmatrix}$$

is positive semidefinite, then also $\mathbf{q} = \mathbf{0}$.

2.3 Show that the outer product Cholesky factorization can also be used to test whether a matrix $M \in \mathbb{R}^{n \times n}$ is positive semidefinite.

2.4 A *rank-constrained* semidefinite program is a problem of the form

$$\begin{aligned}
\text{Maximize} \quad & C \bullet X \\
\text{subject to} \quad & A(X) = \mathbf{b} \\
& X \succeq 0 \\
& \mathrm{rank}(X) \leq k,
\end{aligned}$$

where k is a fixed integer. Show that the problem of solving a rank-constrained semidefinite program is NP-hard for $k = 1$.

2.5 A matrix $M \in \mathbb{R}^{n \times n}$ is called a *Euclidean distance matrix* if there exist n points $\mathbf{p}_1, \ldots, \mathbf{p}_n \in \mathbb{R}^n$, such that M is the matrix of pairwise squared Euclidean distances, i.e.,

$$m_{ij} = \|\mathbf{p}_i - \mathbf{p}_j\|^2, \quad 1 \le i, j \le m.$$

Prove that a matrix M is a Euclidean distance matrix if and only if M is symmetric, $m_{ii} = 0$ for all i, and

$$\mathbf{x}^T M \mathbf{x} \le 0 \ \text{ for all } \mathbf{x} \text{ with } \sum_{i=1}^{n} x_i = 0.$$

2.6 Let $G = (\{1, \ldots, n\}, E)$ be a graph with two edge weight functions $\alpha_e \le \beta_e$, $e \in E$. We want to know whether there exist points $\mathbf{p}_1, \ldots, \mathbf{p}_n \in \mathbb{R}^n$, such that

$$\alpha_{\{i,j\}} \le \|\mathbf{p}_i - \mathbf{p}_j\|^2 \le \beta_{\{i,j\}}, \text{ for all } \{i,j\} \in E.$$

Show that this decision problem can be formulated as a semidefinite program!

2.7 (Sums of squares and minimization I)

(a) Let $p(x) \in \mathbb{R}[x]$ be a univariate polynomial of degree d with real coefficients. We would like to decide whether $p(x)$ is a *sum of squares*, i.e., if it can be written as $p(x) = q_1(x)^2 + \cdots + q_m(x)^2$ for some $q_1(x), \ldots, q_m(x) \in \mathbb{R}[x]$. Formulate this problem as the feasibility of a semidefinite program.

(b) Let us call a polynomial $p(x) \in \mathbb{R}[x]$ *nonnegative* if $p(x) \ge 0$ for all $x \in \mathbb{R}$. Obviously, a sum of squares is nonnegative. Prove that the converse holds as well: Every nonnegative univariate polynomial is a sum of squares. (Hint: First factor into quadratic polynomials.)

(c) Let $p(x) \in \mathbb{R}[x]$ be a given polynomial. Express its global minimum $\min\{p(t) : t \in \mathbb{R}\}$ as the optimum of a suitable semidefinite program (use (b) and a suitable extension of (a)).

2.8 (Sums of squares and minimization II)

(a) Now let $p(x_1, \ldots, x_n)$ be a polynomial in n variables of degree d with real coefficients, and as in Exercise 2.7, we ask whether it can be expressed as a sum of squares (of n-variate real polynomials). Formulate this problem as the feasibility of a semidefinite program. How many variables and constraints are there in this SDP?

(b) Verify that the *Motzkin polynomial* $p(x, y) = 1 + x^2 y^2 (x^2 + y^2 - 3)$ is nonnegative for all pairs $(x, y) \in \mathbb{R}^2$, but it is not a sum of squares.

Even though part (b) shows that for multivariate polynomials, nonnegativity is not equivalent to being a sum of squares, the multivariate version of the method from Exercise 2.7 constitutes a powerful tool in practice, which can find a global minimum in many cases (see, e.g., [Par06] or [Las10]).

Chapter 3
Shannon Capacity and Lovász Theta

Here we will discuss a remarkable geometrically defined graph parameter $\vartheta(G)$. This parameter can be regarded as a semidefinite relaxation of the independence number $\alpha(G)$ of the graph, and also in a dual view, as a semidefinite relaxation of $\chi(\overline{G})$, the chromatic number of G's complement.

Perhaps even more remarkably, $\vartheta(G)$ was invented well *before* the semidefinite era; it predates the Goemans–Williamson algorithm by more than fifteen years and the ideas connected with it contributed very much to the foundations of combinatorial applications of semidefinite programming.

We begin our treatment with presenting Lovász' classical application of $\vartheta(G)$ for determining the Shannon capacity of the 5-cycle. While this material may be well known to many readers, and while for us it presents a detour from the main focus on SDP-based approximation algorithms, we feel that something so impressive and beautiful just cannot be omitted.

3.1 The Similarity-Free Dictionary Problem

Suppose that you have just bought an optical character recognition system. Your goal as a citizen of the internet age may be to digitize all your books so that you can safely throw them away.

However, the system is not perfect and it sometimes gets letters wrong. For example, the letter **E** might mistakenly be recognized as an **F**.

In general, there are *input letters* (the ones in the book) and *output letters* (the ones being recognized). Input and output letters may come from the same alphabet, but also from different ones. Two input letters v, v' are called *similar* if there is an output letter w such that both v and v' may be recognized as w; see Fig. 3.1 for an example.

In the example, $v = $ **E** and $v' = $ **F** are similar, with $w = $ **F** witnessing their similarity. The letters **I**, **J** and **L** are also pairwise similar since all three could be recognized as **I**. Finally, each letter is similar to itself by definition.

B. Gärtner and J. Matoušek, *Approximation Algorithms and Semidefinite Programming*, DOI 10.1007/978-3-642-22015-9_3, © Springer-Verlag Berlin Heidelberg 2012

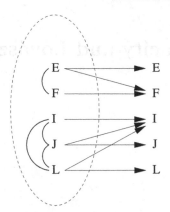

Fig. 3.1 The similarity graph (*left*) connects two input letters if they may be recognized as the same output letter

We can record this information in an (undirected) *similarity* graph that connects two distinct input letters if they are similar; see Fig. 3.1. The information that every letter is similar to itself is implicit.

If the similarity graph is empty, the system can correctly scan all your books: for every recognized output letter w, there is exactly one matching input letter v, and assuming that the system knows its recognition behavior, the correct input letter v can be reconstructed.

But already with a relatively sparse but nonempty similarity graph, the system may get a lot of words wrong. For example, a word with many **E**'s is pretty likely to get corrupted since it suffices for only one of the **E**'s to be mistakenly recognized as an **F**. Such errors can be corrected only if no two distinct words may get recognized as the same word. Given the similarity graph in Fig. 3.1, both **JILL** and **LILI** might turn out as **IIII**, meaning that a book featuring these two characters cannot properly be handled.

Formally, two k-letter words $v_1 \ldots v_k$ and $v'_1 \ldots v'_k$ are called similar if v_i is similar to v'_i for all i. A set of pairwise non-similar words is called a *similarity-free dictionary*.

If the set of input words forms a similarity-free dictionary, then error correction indeed works, since for every recognized word $w_1 \ldots w_k$, there is exactly one word $v_1 \ldots v_k$ in the dictionary such that v_i may be recognized as w_i for all i, and this word must be the correct input word.

While you are waiting for your next book to be scanned, your mind is drifting off and you start asking a theoretical question. What is the largest similarity-free dictionary of k-letter words?

For $k = 1$ (the words are just letters), this is easy to answer: The dictionary must be an *independent set* in the similarity graph. The largest similarity-free dictionary of 1-letter words is therefore a maximum independent set in the similarity graph.

For $k > 1$, we can easily form a graph G^k whose edges characterize similarity between k-letter words. The vertices of G^k are the words of length k, and there is an edge between two words $v_1 \ldots v_k$ and $v'_1 \ldots v'_k$ if they are similar (meaning that v_i is similar to v'_i for all i). This leads to the following:

3.1.1 Observation. *Let $\alpha(G)$ denote the independence number of a graph G, i.e., the size of a maximum independent set in G. Then the largest similarity-free dictionary of k-letter words has size $\alpha(G^k)$.*

It is known that the independence number of a graph is NP-hard to compute [GJ79, Sect. 3.1.3], so finding the size of the largest similarity-free dictionary is hard even for 1-letter words. However, this is not our main concern here, since we want to study the sequence $(\alpha(G^k))_{k \in \mathbb{N}}$ in its entirety. We start by showing that the sequence is *super-multiplicative*.

3.1.2 Lemma. *For all $k, \ell \in \mathbb{N}$,*

$$\alpha(G^{k+\ell}) \geq \alpha(G^k)\alpha(G^\ell).$$

Proof. If I is an independent set in G^k, and J is an independent set in G^ℓ, then the set of $|I||J|$ words

$$\{v_1 \ldots v_k w_1 \ldots w_\ell : v_1 \ldots v_k \in I, w_1 \ldots w_\ell \in J\}$$

is independent in $G^{k+\ell}$. Indeed, no two distinct words in this set can be similar, since this would imply that at least one of I and J contains two distinct similar words. If $|I| = \alpha(G^k)$ and $|J| = \alpha(G^\ell)$, the statement follows.
□

The inequality in Lemma 3.1.2 can be strict. For the 5-cycle C_5 we have $\alpha(C_5) = 2$. But $\alpha(C_5^2) \geq 5 > \alpha(C_5)^2$. To see that $\alpha(C_5^2) \geq 5$, we use the interpretation of $\alpha(C_5^2)$ as the size of a largest similarity-free dictionary of 2-letter words. Suppose that the letters around the cycle C_5 are **A**, **B**, **C**, **D**, **E**. Then it is easy to check that the following five 2-letter words are pairwise non-similar: **AA**, **BC**, **CE**, **DB**, and **ED**. This example is actually the best possible and $\alpha(C_5^2) = 5$.

3.2 The Shannon Capacity

We may view a dictionary as a set of messages, encoded by k-letter words. The goal is to safely transmit any given message over a noisy channel whose input/output behavior induces a similarity graph G as in Fig. 3.1. If the dictionary is similarity-free w.r.t. G, we can indeed correct all errors made during transmission.

Using 1-letter words, we can thus safely transmit $\alpha(G)$ different messages. This means that every letter carries (roughly) $\log(\alpha(G))$ bits of information (the logarithm is binary). Using k-letter words, we can transmit $\alpha(G^k)$ different messages, meaning that each of the k letters carries

$$\frac{1}{k}\log\alpha(G^k)$$

bits of information on the average.

We are interested in the "best" k, the one that leads to the highest information-per-letter ratio. It easily follows from our above considerations on C_5 that $k = 1$ is not always the best choice. Indeed, we have

$$\log\alpha(C_5) = 1 < \frac{1}{2}\log\alpha(C_5^2) = \frac{1}{2}\log 5 \approx 1.161.$$

Consequently, let us define the *Shannon capacity* of a graph G as

$$\sigma(G) = \sup\left\{\frac{1}{k}\log\alpha(G^k) : k \in \mathbb{N}\right\}, \tag{3.1}$$

the (asymptotically) highest information-per-letter ratio that can be achieved. This definition is due to Shannon [Sha56].

3.2.1 Lemma. *For every graph $G = (V, E)$, $\sigma(G)$ is bounded and satisfies*

$$\sigma(G) = \lim_{k\to\infty}\left(\frac{1}{k}\log\alpha(G^k)\right).$$

Proof. Since G^k has $|V|^k$ vertices, we obviously have $\alpha(G^k) \le |V|^k$ which implies that $\sigma(G) \le \log|V|$. Taking logarithms in Lemma 3.1.2, we see that the sequence $(x_k)_{k\in\mathbb{N}} = (\log\alpha(G^k))_{k\in\mathbb{N}}$ is *super-additive*, meaning that $x_{k+\ell} \ge x_k + x_\ell$ for all k, ℓ. Now we use *Fekete's lemma*, which states that for every super-additive sequence $(x_k)_{k\in N}$, the sequence

$$\left(\frac{x_k}{k}\right)_{k\in\mathbb{N}}$$

converges to its supremum (Exercise 3.1 asks you to prove this). □

Shannon already remarked in his original paper [Sha56] in 1956 that it can be quite difficult to compute $\sigma(G)$ even for small graphs G, and in particular he failed to determine $\sigma(C_5)$. We know that

$$\sigma(C_5) \ge \frac{1}{2}\log 5,$$

but it is absolutely not clear whether $k = 2$ yields the best possible information-per-letter ratio.

Only in 1979, could Lovász determine $\sigma(C_5) = \frac{1}{2}\log 5$, showing that the lower bound obtained from 2-letter encodings is tight [Lov79]. Lovász did this by deriving the *theta function*, a new upper bound on $\sigma(G)$ (computable with semidefinite programming, as we will see), and by showing that this upper bound matches the known lower bound for $\sigma(C_5)$.

Instead of $\sigma(G)$, Lovász uses the equivalent quantity

$$\Theta(G) = 2^{\sigma(G)} = \lim_{k\to\infty} \sqrt[k]{\alpha(G^k)} \qquad (3.2)$$

and calls this the Shannon capacity. We will follow his notation. We remark that the Shannon capacity $\Theta(G)$ is bounded from below by $\alpha(G)$ (by super-multiplicativity) and bounded from above by $|V|$. The statement $\sigma(C_5) = \frac{1}{2}\log 5$ now reads as $\Theta(C_5) = \sqrt{5}$.

3.3 The Theta Function

We first pinpoint our earlier notation of similarity. Here and in the following, we assume that a graph with n vertices has the vertex set $\{1, 2, \ldots, n\}$.

3.3.1 Definition. *Let $G = (V, E)$ be a graph. Vertices i and j are called similar in G if either $i = j$ or $\{i, j\} \in E$.*

We remark that the negative statement "i is not similar to j" is more conveniently expressed as "$\{i, j\} \in \overline{E}$," where $\overline{E} := \binom{V}{2} \setminus E$ is the edge set of the *complementary graph* $\overline{G} = (V, \overline{E})$. Here $\binom{V}{2}$ is the set of all the two-element subsets of V.

3.3.2 Definition. *An orthonormal representation of a graph $G = (V, E)$ with n vertices is a sequence $\mathcal{U} = (\mathbf{u}_1, \mathbf{u}_2, \ldots, \mathbf{u}_n)$ of unit vectors in S^{n-1} such that*

$$\mathbf{u}_i^T \mathbf{u}_j = 0 \text{ if } \{i, j\} \in \overline{E}. \qquad (3.3)$$

It is clear that every graph has such a representation, since we may take the n pairwise orthogonal unit vectors $\mathbf{e}_1, \ldots, \mathbf{e}_n$.

But we are looking for a better representation, if possible. Intuitively, a representation is good if its vectors fit into a small spherical cap.

Formally, and at first glance somewhat arbitrarily, we define the *value* of an orthonormal representation $\mathcal{U} = (\mathbf{u}_1, \mathbf{u}_2, \ldots, \mathbf{u}_n)$ as

$$\vartheta(\mathcal{U}) := \min_{\|\mathbf{c}\|=1} \max_{i=1}^{n} \frac{1}{(\mathbf{c}^T \mathbf{u}_i)^2}. \qquad (3.4)$$

The minimum exists, since we can cast the problem as the minimization of a continuous function over a compact set (the unit sphere S^{n-1} minus suitable open sets around the hyperplanes $\{\mathbf{x} : \mathbf{x}^T\mathbf{u}_i = 0\}$ to avoid singularities). A vector $\mathbf{c}$ that attains the minimum is called a *handle* of $\mathcal{U}$.

3.3.3 Definition. *The theta function $\vartheta(G)$ of G is the smallest value $\vartheta(\mathcal{U})$ over all orthonormal representations $\mathcal{U}$ of G.*

Again, the minimum exists, since (i) $\vartheta(\mathcal{U})$ is continuous, and (ii) the set of orthonormal representations is the compact set $(S^{n-1})^n$, intersected with closed sets of the form $\{\mathbf{u}_i^T\mathbf{u}_j = 0\}$ (which again yields a compact set).

Let us illustrate this notion on two extreme examples. If G is the complete graph, then every sequence of n unit vectors is an orthonormal representation, and every $\mathbf{c} \in S^{n-1}$ is a handle for the optimal orthonormal representation given by $\mathbf{u}_i = \mathbf{c}$ for all i. The theta function $\vartheta(G)$, the value of such an optimal representation, is 1. If G is the empty graph, the $\mathbf{u}_i$ must form an orthonormal basis of $\mathbb{R}^n$. We may assume that $\mathbf{u}_i = \mathbf{e}_i$ for all i. Then the handle is $\mathbf{c} = \sum_{i=1}^n \mathbf{u}_i/\sqrt{n}$, resulting in $\mathbf{c}^T\mathbf{u}_i = 1/\sqrt{n}$ for all i, and hence $\vartheta(G) = n$.

3.4 The Lovász Bound

In this section we show that $\vartheta(G)$ is an upper bound for the Shannon capacity $\Theta(G)$. This requires two lemmas.

With the definition of the graph G^k on page 29, we want to prove that $\vartheta(G^k) \leq \vartheta(G)^k$ (recall that the inverse inequality holds for the independence number α, by Lemma 3.1.2). For this, we first handle the case $k = 2$, in the following more general form.

3.4.1 Definition. *Let $G = (V, E)$ and $H = (W, F)$ be graphs. The strong product of G and H is the graph $G \cdot H$ with vertex set $V \times W$, and an edge between (v, w) and (v', w') if v is similar to v' in G and w is similar to w' in H.*

This is called the strong product to distinguish it from the usual graph product $G \times H$ in which there is an edge between (v, w) and (v', w') if $\{v, v'\} \in E$ and $\{w, w'\} \in F$; see Fig. 3.2. In fact, the strong product is obtained from the usual product by adding the edges of the *square product* (or grid) of G and H. In this latter product, there is an edge between (v, w) and (v', w') if they differ in exactly one component, and the two involved vertices form an edge in the respective graph.

3.4.2 Lemma. *For all graphs G and H,*

$$\vartheta(G \cdot H) \leq \vartheta(G)\vartheta(H).$$

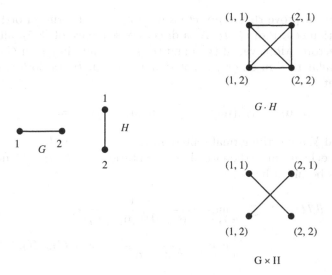

Fig. 3.2 The strong product $G \cdot H$ of two graphs G and H is a supergraph of the product $G \times H$

We remark that the claim of the lemma actually holds with equality; see Exercise 4.13.

Since G^k is isomorphic to the k-fold strong product $(\cdots((G\cdot G)\cdot G)\cdots G)$, we have the following consequence.

3.4.3 Corollary. $\vartheta(G^k) \leq \vartheta(G)^k$.

Proof of Lemma 3.4.2. Let $\mathcal{U} = (\mathbf{u}_1, \mathbf{u}_2, \ldots, \mathbf{u}_m)$ and $\mathcal{V} = (\mathbf{v}_1, \mathbf{v}_2, \ldots, \mathbf{v}_n)$ be optimal orthonormal representations of $G = (V, E)$ and $H = (W, F)$, with handles $\mathbf{c}$ and $\mathbf{d}$. From this we will construct an orthonormal representation of $G \cdot H$ with value at most $\vartheta(G)\vartheta(H)$.

The construction is simple: the orthonormal representation is obtained by taking all *tensor products* of vectors $\mathbf{u}_i$ with vectors $\mathbf{v}_j$, and an upper bound for its value is computed using the tensor product of the handles $\mathbf{c}$ and $\mathbf{d}$.

The tensor product of two vectors $\mathbf{x} \in \mathbb{R}^m$ and $\mathbf{y} \in \mathbb{R}^n$ is the (column) vector $\mathbf{x} \otimes \mathbf{y} \in \mathbb{R}^{mn}$ defined by

$$\mathbf{x} \otimes \mathbf{y} = (x_1y_1, \ldots, x_1y_n, x_2y_1, \ldots, x_2y_n, \ldots, x_my_1, \ldots, x_my_n) \in \mathbb{R}^{mn}.$$

Equivalently, the tensor product is the matrix $\mathbf{x}\mathbf{y}^T$, written as one long vector (row by row). We have the following identity:

$$(\mathbf{x} \otimes \mathbf{y})^T(\mathbf{x}' \otimes \mathbf{y}') = \sum_{i=1}^{m}\sum_{j=1}^{n} x_iy_jx_i'y_j' = \sum_{i=1}^{m} x_ix_i' \sum_{j=1}^{n} y_jy_j' = (\mathbf{x}^T\mathbf{x}')(\mathbf{y}^T\mathbf{y}').$$

$$(3.5)$$

Now we can prove that the vectors $\mathbf{u}_i \otimes \mathbf{v}_j$ indeed form an orthonormal representation $\mathcal{U} \otimes \mathcal{V}$ of $G \cdot H$. As a direct consequence of (3.5), all of them are unit vectors. Moreover, if (i, j) and (i', j') are not similar in $G \cdot H$, then i is not similar to i' in G, or j is not similar to j' in H. In both cases, (3.5) implies that

$$(\mathbf{u}_i \otimes \mathbf{v}_j)^T (\mathbf{u}_{i'} \otimes \mathbf{v}_{j'}) = (\mathbf{u}_i^T \mathbf{u}_{i'})(\mathbf{u}_j^T \mathbf{u}_{j'}) = 0,$$

since $\mathcal{U}$ and $\mathcal{V}$ are orthonormal representations of G and H.

Thus, we have an orthonormal representation of $G \cdot H$. By definition, $\vartheta(G \cdot H)$ is bounded by

$$
\begin{aligned}
\vartheta(\mathcal{U} \otimes \mathcal{V}) &\leq \max_{i \in V, j \in W} \frac{1}{((\mathbf{c} \otimes \mathbf{d})^T (\mathbf{u}_i \otimes \mathbf{v}_j))^2} \\
&= \max_{i \in V, j \in W} \frac{1}{(\mathbf{c}^T \mathbf{u}_i)^2 (\mathbf{d}^T \mathbf{v}_j)^2} = \vartheta(G)\vartheta(H). \qquad \square
\end{aligned}
$$

Here is the second lemma that we need: The theta function $\vartheta(G)$ is an upper bound for the independence number of G.

3.4.4 Lemma. *For every graph G, $\alpha(G) \leq \vartheta(G)$.*

Proof. Let $I \subseteq V(G)$ be a maximum independent set in G, and let $\mathcal{U} = (\mathbf{u}_1, \mathbf{u}_2, \ldots, \mathbf{u}_n)$ be an optimal orthonormal representation of G with handle $\mathbf{c}$.

We know that the vectors $\mathbf{u}_i$, $i \in I$, are pairwise orthogonal, which implies (Exercise 3.3) that

$$\mathbf{c}^T \mathbf{c} \geq \sum_{i \in I} (\mathbf{c}^T \mathbf{u}_i)^2.$$

We thus have

$$1 = \mathbf{c}^T \mathbf{c} \geq \sum_{i \in I} (\mathbf{c}^T \mathbf{u}_i)^2 \geq |I| \min_{i \in I} (\mathbf{c}^T \mathbf{u}_i)^2 = \alpha(G) \min_{i \in I} (\mathbf{c}^T \mathbf{u}_i)^2.$$

This in turn means that

$$\alpha(G) \leq \frac{1}{\min_{i \in I} (\mathbf{c}^T \mathbf{u}_i)^2} = \max_{i \in I} \frac{1}{(\mathbf{c}^T \mathbf{u}_i)^2} \leq \max_{i \in V} \frac{1}{(\mathbf{c}^T \mathbf{u}_i)^2} = \vartheta(G). \qquad \square$$

Now we are ready for the main result of this section.

3.4.5 Theorem (Lovász' bound). *For every graph G, $\Theta(G) \leq \vartheta(G)$.*

Proof. By Lemma 3.4.4 and Corollary 3.4.3, for all k we have

$$\alpha(G^k) \leq \vartheta(G^k) \leq \vartheta(G)^k.$$

It follows that

$$\sqrt[k]{\alpha(G^k)} \leq \vartheta(G),$$

and hence

$$\Theta(G) = \lim_{k \to \infty} \sqrt[k]{\alpha(G^k)} \leq \vartheta(G). \qquad \square$$

3.5 The 5-Cycle

Using the bound of Theorem 3.4.5, we can now determine the Shannon capacity $\Theta(C_5)$ of the 5-cycle. We already know that $\Theta(C_5) \geq \sqrt{5}$, by using 2-letter encodings. The fact that this is optimal follows from the next lemma, together with Theorem 3.4.5.

3.5.1 Lemma. $\vartheta(C_5) \leq \sqrt{5}$.

Proof. We need to find an orthonormal representation of C_5 with value at most $\sqrt{5}$. Let the vertices of C_5 be $1, 2, 3, 4, 5$ in cyclic order.

Here is Lovász' "umbrella construction" that yields vectors $\mathbf{u}_1, \dots, \mathbf{u}_5$ in S^2 (we can add two zero coordinates to lift them into S^4). Imagine an umbrella with unit handle $\mathbf{c} = (0, 0, 1)$ and five unit ribs of the form

$$\mathbf{u}_i = \frac{(\cos \frac{2\pi i}{5}, \sin \frac{2\pi i}{5}, z)}{\|(\cos \frac{2\pi i}{5}, \sin \frac{2\pi i}{5}, z)\|}, \quad i = 1, \dots, 5.$$

If $z = 0$, the umbrella is completely "flat," as in Fig. 3.3 left (this is a top view where $\mathbf{c}$ collapses to the origin). Letting z grow to ∞ corresponds to the process of folding up the umbrella.

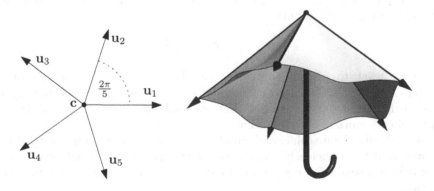

Fig. 3.3 A five-rib umbrella: fully open and viewed from the top (*left*), and partially folded so that non-adjacent ribs are perpendicular (*right*)

We keep folding the umbrella until the vectors $\mathbf{u}_5$ and $\mathbf{u}_2$ become orthogonal. This will eventually happen since we start with angle $4\pi/5 > \pi/2$ in the flat position and converge to angle 0 as $z \to \infty$. We can compute the value of z for which we get orthogonality: We must have

$$0 = \mathbf{u}_5^T \mathbf{u}_2 \quad \Leftrightarrow \quad (1,0,z) \begin{pmatrix} \cos\frac{4\pi}{5} \\ \sin\frac{4\pi}{5} \\ z \end{pmatrix} = \cos\frac{4\pi}{5} + z^2 = 0.$$

Hence

$$z = \sqrt{-\cos\frac{4\pi}{5}}, \quad \mathbf{u}_5 = \frac{\left(1, 0, \sqrt{-\cos\frac{4\pi}{5}}\right)}{\sqrt{1 - \cos\frac{4\pi}{5}}}.$$

For this value of z, symmetry implies that we do have an orthonormal representation $\mathcal{U}$ of C_5: every $\mathbf{u}_i$ is orthogonal to the two "opposite" vectors $\mathbf{u}_{i+2}$ and $\mathbf{u}_{i-2}$ of its two non-neighbors in C_5 (indices wrap around). Recalling that $\mathbf{c} = (0,0,1)$, we have

$$\vartheta(C_5) \le \vartheta(\mathcal{U}) \le \max_{i=1}^{5} \frac{1}{(\mathbf{c}^T \mathbf{u}_i)^2} = \frac{1}{(\mathbf{c}^T \mathbf{u}_5)^2} = \frac{1 - \cos\frac{4\pi}{5}}{-\cos\frac{4\pi}{5}},$$

by symmetry. Exercise 3.4 asks you to prove that this number equals $\sqrt{5}$.

$\square$

3.6 Two Semidefinite Programs for the Theta Function

The value of $\Theta(C_5)$ was unknown for more than 20 years after Shannon had given the lower bound $\Theta(C_5) \ge \sqrt{5}$. Together with the Lovász bounds $\Theta(C_5) \le \vartheta(C_5) \le \sqrt{5}$, we get

$$\Theta(C_5) = \vartheta(C_5) = \sqrt{5}.$$

Here we want to discuss how $\vartheta(G)$ can be computed for an arbitrary graph $G = (V, E)$. The above method for C_5 was somewhat ad-hoc, and only in hindsight did it turn out that the umbrella construction yields an optimal orthonormal representation.

Armed with the machinery of semidefinite programming, it is not hard to show that $\vartheta(G)$ is efficiently computable (with an arbitrarily small error). We can obtain a semidefinite program for it by more or less just rewriting the definition.

We recall that $\vartheta(G)$ is the smallest value of

$$\vartheta(\mathcal{U}) = \min_{\|\mathbf{c}\|=1} \max_{i \in V} \frac{1}{(\mathbf{c}^T \mathbf{u}_i)^2},$$

over all orthonormal representations $\mathcal{U}$. By replacing $\mathbf{u}_i$ with $-\mathbf{u}_i$ if necessary, we may assume $\mathbf{c}^T \mathbf{u}_i \geq 0$ for all i. But then

$$\frac{1}{\sqrt{\vartheta(G)}} = \max_{\mathcal{U}} \frac{1}{\sqrt{\vartheta(\mathcal{U})}} = \max_{\mathcal{U}} \max_{\|\mathbf{c}\|=1} \min_{i \in V} \mathbf{c}^T \mathbf{u}_i.$$

We introduce an additional variable $t \in \mathbb{R}_+^n$ representing the minimum, and then we get $1/\sqrt{\vartheta(G)}$ as the value of the vector program

$$
\begin{aligned}
\text{maximize} \quad & t \\
\text{subject to} \quad & \mathbf{u}_i^T \mathbf{u}_j = 0 \text{ for all } \{i,j\} \in \overline{E} \\
& \mathbf{c}^T \mathbf{u}_i \geq t, \ i \in V \\
& \|\mathbf{u}_i\| = 1, \ i \in V \\
& \|\mathbf{c}\| = 1.
\end{aligned}
\tag{3.6}
$$

This does not yet have the form of a semidefinite program in equational form, but it can be brought into this form; see Exercise 3.2 and Sect. 2.5.

There are several interesting ways of expressing $\vartheta(G)$ itself as a value of a suitable semidefinite program (above we expressed $1/\sqrt{\vartheta(G)}$ rather than $\vartheta(G)$). Proving that one really obtains $\vartheta(G)$ from these semidefinite programs is not immediate (although not too hard either).

Here we will present just one of these semidefinite formulations (several others can be found, e.g., in Knuth [Knu94]). In this particular case our work in the proof of the next theorem will pay off in the next section, where we relate $\vartheta(G)$ to the chromatic number of G's complement.

3.6.1 Theorem. *For every graph $G = (V, E)$ with $V = \{1, \ldots, n\}$, the theta function $\vartheta(G)$ is the value of the following semidefinite program in the matrix variable $Y \in \mathrm{SYM}_n$ and the real variable t.*

$$
\begin{aligned}
\text{Minimize} \quad & t \\
\text{subject to} \quad & y_{ij} = -1 && \text{if } \{i,j\} \in \overline{E} \\
& y_{ii} = t - 1 && \text{for all } i = 1, \ldots, n \\
& Y \succeq 0.
\end{aligned}
\tag{3.7}
$$

Proof. Let us denote the value of (3.7) by $\vartheta'(G)$. We first show that $\vartheta'(G) \leq \vartheta(G)$. Let $\mathcal{U} = (\mathbf{u}_1, \mathbf{u}_2, \ldots, \mathbf{u}_n)$ be an optimal orthonormal representation of G with handle $\mathbf{c}$. We define a matrix $\tilde{Y} \in \mathrm{SYM}_n$ by

$$\tilde{y}_{ij} := \frac{\mathbf{u}_i^T \mathbf{u}_j}{(\mathbf{c}^T \mathbf{u}_i)(\mathbf{c}^T \mathbf{u}_j)} - 1, \quad i \neq j$$

and

$$\tilde{y}_{ii} := \vartheta(G) - 1, \quad i = 1, \ldots, n.$$

Since $\mathcal{U}$ is an orthonormal representation, we have $\tilde{y}_{ij} = -1$ for $\{i, j\} \in \overline{E}$. If we can show that $\hat{Y} \succeq 0$, we know that the pair $(\tilde{Y}, \tilde{t})$ with $\tilde{t} = \vartheta(G)$) is a feasible solution of (3.7), meaning that the program's value $\vartheta'(G)$ is at most $\vartheta(G)$.

To see $\tilde{Y} \succeq 0$, we first observe (a simple calculation) that

$$\tilde{y}_{ij} = \left(\mathbf{c} - \frac{\mathbf{u}_i}{\mathbf{c}^T \mathbf{u}_i}\right)^T \left(\mathbf{c} - \frac{\mathbf{u}_j}{\mathbf{c}^T \mathbf{u}_j}\right), \quad i \neq j,$$

and (by definition of $\vartheta(G)$)

$$\tilde{y}_{ii} = \vartheta(G) - 1 \geq \frac{1}{(\mathbf{c}^T \mathbf{u}_i)^2} - 1 = \left(\mathbf{c} - \frac{\mathbf{u}_i}{\mathbf{c}^T \mathbf{u}_i}\right)^T \left(\mathbf{c} - \frac{\mathbf{u}_i}{\mathbf{c}^T \mathbf{u}_i}\right).$$

This means that $\tilde{Y}$ is of the form $\tilde{Y} = D + U^T U$, where D is a diagonal matrix with nonnegative entries, and U is the matrix whose i-th column is the vector $\mathbf{c} - \mathbf{u}_i / \mathbf{c}^T \mathbf{u}_i$. Thus, $\tilde{Y} \succeq 0$.

To show that the value of (3.7) is at least $\vartheta(G)$, we let $(\tilde{Y}, \tilde{t})$ be any feasible solution of (3.7) with the property that $\tilde{t}$ is minimal subject to $\tilde{y}_{ij}$ fixed for $i \neq j$. This implies that $\tilde{Y}$ has one eigenvalue equal to 0 (otherwise, we could decrease all $\tilde{y}_{ii}$ and $\tilde{t}$) and it is therefore singular. Note that $\tilde{t} \geq 1$.

Now let $\tilde{Y} = S^T S$ be a Cholesky factorization of $\tilde{Y}$; see Fact 2.2.1(iii). Let $\mathbf{s}_1, \ldots, \mathbf{s}_n$ be the columns of S. Since $\tilde{Y}$ is singular, S is singular as well, and the $\mathbf{s}_i$ span a proper subspace of $\mathbb{R}^n$. Consequently, there exists a unit vector $\mathbf{c}$ that is orthogonal to all the $\mathbf{s}_i$.

Next we define

$$\mathbf{u}_i := \frac{1}{\sqrt{\tilde{t}}}(\mathbf{c} + \mathbf{s}_i), \quad i = 1, \ldots, n,$$

and we intend to show that $\mathcal{U} = \{\mathbf{u}_1, \ldots, \mathbf{u}_n\}$ is an orthonormal representation of G. For this, we compute

$$\mathbf{u}_i^T \mathbf{u}_j = \frac{1}{\tilde{t}}(\mathbf{c} + \mathbf{s}_i)^T (\mathbf{c} + \mathbf{s}_j) = \frac{1}{\tilde{t}}(\underbrace{\mathbf{c}^T \mathbf{c}}_{1} + \underbrace{\mathbf{c}^T \mathbf{s}_j}_{0} + \underbrace{\mathbf{s}_i^T \mathbf{c}}_{0} + \underbrace{\mathbf{s}_i^T \mathbf{s}_j}_{\tilde{y}_{ij}}) = \frac{1}{\tilde{t}}(1 + \tilde{y}_{ij}).$$

Since $\tilde{y}_{ii} = \tilde{t} - 1$, we get

$$\|\mathbf{u}_i\|^2 = \mathbf{u}_i^T \mathbf{u}_i = \frac{1}{\tilde{t}}(1 + \tilde{t} - 1) = 1.$$

Similarly, if $\{i, j\} \in \overline{E}$, then $\tilde{y}_{ij} = -1$, and so

$$\mathbf{u}_i^T \mathbf{u}_j = \frac{1}{\tilde{t}}(1 - 1) = 0.$$

So we have indeed found an orthonormal representation of G. Since we further have

$$(\mathbf{c}^T \mathbf{u}_i)^2 = \left(\mathbf{c}^T \frac{1}{\sqrt{\tilde{t}}}(\mathbf{c} + \mathbf{s}_i)\right)^2 = \frac{1}{\tilde{t}}\left(\mathbf{c}^T(\mathbf{c} + \mathbf{s}_i)\right)^2 = \frac{1}{\tilde{t}}, \quad i = 1, \ldots, n,$$

we get

$$\vartheta(G) \leq \vartheta(\mathcal{U}) \leq \max_{i=1}^{n} \frac{1}{(\mathbf{c}^T \mathbf{u}_i)^2} = \tilde{t},$$

which completes the proof, since we may choose $\tilde{t}$ arbitrarily close to $\vartheta'(G)$.

$\square$

3.7 The Sandwich Theorem and Perfect Graphs

We know that $\vartheta(G)$ is bounded below by $\alpha(G)$, the independence number of the graph G. But we can also bound $\vartheta(G)$ from above in terms of another graph parameter. This bound will also shed more light on the geometric interpretation of the semidefinite program (3.7) for $\vartheta(G)$.

3.7.1 Definition. *Let $G = (V, E)$ be a graph.*

(i) *A clique in G is a subset $K \subseteq V$ of vertices such that $\{v, w\} \in E$ for all distinct $v, w \in K$. The clique number $\omega(G)$ of G is the size of a largest clique in G.*

(ii) *A proper k-coloring of G is a mapping $c\colon V \to \{1, \ldots, k\}$ such that $c(v) \neq c(w)$ if $\{v, w\} \in E$. The chromatic number $\chi(G)$ of G is the smallest k such that G has a proper k-coloring.*

According to this definition, an independent set in G is a clique in the complementary graph $\overline{G}$, and vice versa. Consequently,

$$\alpha(G) = \omega(\overline{G}). \tag{3.8}$$

Here is the promised upper bound on $\vartheta(G)$. Together with the already known lower bound, we obtain the *Sandwich Theorem* that bounds $\vartheta(G)$ in terms of clique number and chromatic number of the complementary graph.

3.7.2 Theorem. *For every graph $G = (V, E)$,*

$$\omega(\overline{G}) \leq \vartheta(G) \leq \chi(\overline{G}).$$

Proof. The lower bound on $\vartheta(G)$ is immediate from Lemma 3.4.4 and (3.8). For the upper bound, let us suppose that $\vartheta(G) > 1$ (the bound is trivial for

$\vartheta(G) = 1$). But then $\chi(\overline{G}) \geq 2$, since a 1-coloring is possible only for $\overline{E} = \emptyset$, in which case $\vartheta(G) = 1$.

Now let us rescale (3.7) into the following equivalent form (as usual, we assume that $V = \{1, \ldots, n\}$):

$$
\begin{aligned}
\text{Minimize} \quad & t \\
\text{subject to} \quad & y_{ij} = -1/(t-1) \quad \text{for all } \{i,j\} \in \overline{E} \\
& y_{ii} = 1 \quad \text{for all } i = 1, \ldots, n \\
& Y \succeq 0.
\end{aligned}
\tag{3.9}
$$

If we rewrite $Y \succeq 0$ as $Y = S^T S$ for S a matrix with columns $\mathbf{s}_1, \ldots, \mathbf{s}_n$, the equality constraints of (3.9) translate as follows:

$$
y_{ij} = -\frac{1}{t-1} \quad \Leftrightarrow \quad \mathbf{s}_i^T \mathbf{s}_j = -\frac{1}{t-1}
$$

$$
y_{ii} = 1 \quad \Leftrightarrow \quad \|\mathbf{s}_i\| = 1.
$$

Lemma 3.7.4 below shows that if $\overline{G}$ has a proper k-coloring, then we can find vectors $\mathbf{s}_i$ that satisfy the latter equations with $t = k$. This implies that $(Y, t) = (S^T S, k)$ is a feasible solution of (3.9), and hence $k \geq \vartheta(G)$, the value of (3.9). The upper bound follows if we choose $k = \chi(\overline{G})$. $\qquad\square$

The vectors $\mathbf{s}_i$ constructed in this proof can be regarded as a "coloring" of $\overline{G}$ by vectors. This motivates the following definition.

3.7.3 Definition. *For* $k \in \mathbb{R}$, *a* vector k-coloring[1] *of a graph* $G = (V, E)$ *is a mapping* $\gamma: V \to S^{n-1}$ *such that*

$$
\gamma(v)^T \gamma(w) = -\frac{1}{k-1}, \quad \{v, w\} \in E.
$$

For a (proper) k-coloring, we require that adjacent vertices have different colors. For a vector k-coloring, we require the "colors" of adjacent vertices to have a large angle. The proof of Theorem 3.7.2 shows that $\vartheta(G)$ is the smallest k such that $\overline{G}$ has a vector k-coloring. The upper bound $\vartheta(G) \leq \chi(\overline{G})$ then follows from the fact that the notion of vector k-colorings is a relaxation of the notion of proper k-colorings:

3.7.4 Lemma. *If a graph* G *has a* k-coloring, *then it also has a vector* k-coloring.

Proof. We construct k unit-length vectors $\mathbf{u}_1, \ldots, \mathbf{u}_k$ such that

$$
\mathbf{u}_i^T \mathbf{u}_j = -\frac{1}{k-1}, \quad i \neq j.
$$

---

[1] This notion is sometimes called *strict* vector k-coloring to distinguish it from a *non-strict* k-vector coloring, where the requirement $\gamma(v)^T \gamma(w) = -\frac{1}{k-1}$ is weakened to $\gamma(v)^T \gamma(w) \leq -\frac{1}{k-1}$. In this chapter we deal exclusively with strict vector colorings.

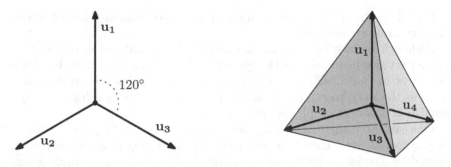

Fig. 3.4 Unit vectors with pairwise scalar products $-1/(k-1)$ for $k = 3, 4$

Given a k-coloring c of G, a vector k-coloring of G can then be obtained by setting $\gamma(v) := \mathbf{u}_{c(v)}$, $v \in V$. The k vectors form the vertices of a regular simplex centered at the origin; see Fig. 3.4 for the cases $k = 3, 4$. In general, we define

$$\mathbf{u}_i = \frac{\mathbf{e}_i - \frac{1}{k}\sum_{\ell=1}^{k}\mathbf{e}_\ell}{\|\mathbf{e}_i - \frac{1}{k}\sum_{\ell=1}^{k}\mathbf{e}_\ell\|}, \quad i = 1, \dots, k. \qquad \square$$

Perfect graphs. We know that the clique number $\omega(G)$ is NP-hard to compute for general graphs. The same can be said about the chromatic number $\chi(G)$. But there is a class of graphs for which Theorem 3.7.2 makes both values computable in polynomial time.

3.7.5 Definition. *A graph G is called* perfect *if $\omega(G') = \chi(G')$ for every induced subgraph G' of G.*

There are many known families of perfect graphs. The perhaps simplest nontrivial examples are bipartite graphs. Indeed, every induced subgraph of a bipartite graph is again bipartite, and every bipartite graph has clique number and chromatic number 2.

Other examples of perfect graphs are *interval graphs* (intersection graphs of closed intervals on the real line), and more generally, *chordal graphs* (every cycle of length at least four has an edge connecting two vertices that are not neighbors along the cycle).

For perfect graphs, Theorem 3.7.2 implies

$$\omega(G) = \vartheta(\overline{G}) = \chi(G),$$

meaning that maximum cliques and minimum colorings can be computed for perfect graphs in polynomial time through semidefinite programming. Indeed, since we are looking for an integer, it suffices to solve (3.9) (for the complementary graph) up to accuracy $\varepsilon < 1/2$. Moreover, due to $y_{ii} = 1$, all entries of a feasible Y are scalar products of unit vectors and hence in

$[-1, 1]$. This means that our requirements for polynomial-time solvability (see Sect. 2.6) are satisfied.

Although remarkable progress has recently been achieved in understanding perfect graphs (a proof of the *Strong Perfect Graph Conjecture* by Chudnovsky et al. [CRST06]), the approach based on semidefinite programming remains the *only* known polynomial-time method for computing the clique number of a perfect graph (as far as we know).

One can also compute the independence number $\alpha(G)$ of a perfect graph G in polynomial time. Indeed, according to the *weak perfect graph conjecture*, proved by Lovász in 1972, the complement of every perfect graph is also perfect. So $\omega(\overline{G})$ is polynomial-time computable by the above, and it equals $\alpha(G)$ by definition.

Exercises

3.1 Let $(x_k)_{k \in \mathbb{N}}$ be a sequence of real numbers such that for all natural numbers k and ℓ,

$$x_{k+\ell} \geq x_k + x_\ell.$$

We say that the sequence is *super-additive*. Prove that

$$\lim_{k \to \infty} \frac{x_k}{k} = \sup\{\frac{x_k}{k} : k \in \mathbb{N}\},$$

where both the limit and the supremum may be ∞.

3.2 Prove that the program (3.6) can be rewritten into a semidefinite program in equational form, and with the same value.

3.3 Let $\mathbf{u}_1, \ldots, \mathbf{u}_k$ be pairwise orthogonal unit vectors in $\mathbb{R}^n$. Prove that $\mathbf{c}^T \mathbf{c} \geq \sum_{i=1}^{k} (\mathbf{c}^T \mathbf{u}_i)^2$ for all $\mathbf{c} \in \mathbb{R}^n$.

3.4 Prove that

$$\frac{1 - \cos \frac{4\pi}{5}}{-\cos \frac{4\pi}{5}} = \sqrt{5}.$$

3.5 For graphs G, H, let $G + H$ stand for the disjoint union of G and H. Formally, we let H' be an isomorphic copy of H whose vertex set is disjoint from $V(G)$, and we put $V(G + H) := V(G) \cup V(H')$, $E(G + H) := E(G) \cup E(H')$. Prove that $\Theta(G + H) \geq \Theta(G) + \Theta(H)$.

3.6 With the definition of $G + H$ as in Exercise 3.5, prove that $\vartheta(G + \overline{G}) \geq \sqrt{2|V(G)|}$, where $\overline{G}$ stands for the complement of G.

3.7 Let $G = (V, E)$ be a graph, and let $\mathbb{K}$ be a field (such as the reals, the complex numbers, or a finite field $\mathrm{GF}(q)$). Let a *functional representation* $\mathcal{F}$ of G over $\mathbb{K}$ consist of the following:

1. A ground set X (an arbitrary set, not necessarily related to G or $\mathbb{K}$ in any way)
2. An element $c_v \in X$ for every vertex $v \in V$
3. A function $f_v : X \to \mathbb{K}$ for every vertex $v \in V$

These objects have to satisfy

(i) $f_v(c_v) \neq 0$ for every $v \in V$
(ii) If $\{u, v\} \in \overline{E}$, then $f_u(c_v) = 0$

We write $\mathcal{F} = (X, (c_v, f_v)_{v \in V})$. The *dimension* $\dim \mathcal{F}$ of $\mathcal{F}$ is the dimension of the subspace generated by all the functions f_v, $v \in V$, in the vector space $\mathbb{K}^X$ of all functions $X \to \mathbb{K}$.

(a) Check that $\alpha(G) \leq \dim \mathcal{F}$ for every functional representation $\mathcal{F}$ of G (over any field).
(b) In what sense can an orthonormal representation $\mathcal{U}$ of G be regarded as a functional representation of G over $\mathbb{R}$? Compare the strength of the resulting upper bounds on $\alpha(G)$.
(c) Suppose that a graph $G = (V, E)$ has a functional representation $\mathcal{F}$ over some field $\mathbb{K}$ and that $G' = (V', E')$ has a functional representation $\mathcal{F}'$ over the same $\mathbb{K}$. Then the strong product $G \cdot G'$ has a functional representation over $\mathbb{K}$ of dimension at most $(\dim \mathcal{F})(\dim \mathcal{F}')$. Infer that $\Theta(G) \leq \dim(\mathcal{F})$ for every functional representation of G.

Remark: For suitable graphs G and a suitable choice of the field $\mathbb{K}$, the upper bound in (c) on the Shannon capacity can be smaller than $\vartheta(G)$. This tool was developed and used by Alon [Alo98] in surprising results on the Shannon capacity.

Chapter 4
Duality and Cone Programming

4.1 Introduction

One of the most important results in linear programming is arguably the *duality theorem*. Semidefinite programming also has a duality theorem, but its formulation and proof are less straightforward than in the case of linear programming.

Instead of developing the duality theorem for semidefinite programming directly, we will work in the more general setting of *cone programming*. This abstraction allows us to see the essence more clearly and illustrate it with simple geometric examples. Moreover, it can be useful in other contexts; see Chap. 7.

According to Definition 2.4.1, a semidefinite program in equational form is an optimization problem of the form

$$
\begin{aligned}
\text{maximize} \quad & C \bullet X \\
\text{subject to} \quad & A_i \bullet X = b_i, \quad i = 1, 2, \ldots, m \\
& X \succeq 0.
\end{aligned}
\tag{4.1}
$$

Here X is an unknown $n \times n$ symmetric real matrix (which we write as $X \in \mathrm{SYM}_n$). The input data are a matrix $C \in \mathrm{SYM}_n$ specifying the objective function, a vector $\mathbf{b} \in \mathbb{R}^m$, and matrices $A_i \in \mathrm{SYM}_n$, $i = 1, 2, \ldots, m$.

For the purposes of this chapter, we summarize the m equality constraints in the form $A(X) = \mathbf{b}$, where $A \colon \mathrm{SYM}_n \to \mathbb{R}^m$ is the linear mapping

$$
A(X) = (A_1 \bullet X, A_2 \bullet X, \ldots, A_m \bullet X).
$$

Instead of a linear mapping, we will often use the term *linear operator* in this context.

The main goal of this chapter is to derive the following *strong duality theorem of semidefinite programming*.

B. Gärtner and J. Matoušek, *Approximation Algorithms and Semidefinite Programming*, DOI 10.1007/978-3-642-22015-9_4,
© Springer-Verlag Berlin Heidelberg 2012

4.1.1 Theorem. *If the semidefinite program (4.1) is feasible and has a finite value γ, and if there is a positive definite matrix $\tilde{X}$ such that $A(\tilde{X}) = \mathbf{b}$, then the dual program*

$$
\begin{aligned}
&minimize \quad \mathbf{b}^T \mathbf{y} \\
&subject\ to \quad \textstyle\sum_{i=1}^{m} y_i A_i - C \succeq 0
\end{aligned}
\tag{4.2}
$$

is feasible and has finite value $\beta = \gamma$.

Having a positive definite (as opposed to merely a positive semidefinite) matrix $\tilde{X}$ such that $A(\tilde{X}) = \mathbf{b}$ is also referred to as "strict feasibility."

The link to cone programming is established by the fact that the set $\mathrm{PSD}_n = \{X \in \mathrm{SYM}_n : X \succeq 0\}$ of positive semidefinite matrices is a *closed convex cone*.

The outline of this chapter is as follows. We first define closed convex cones and their duals. We prove a simple but powerful separation theorem for closed convex cones that can already be considered as a very basic duality theorem. Based on this, and bringing in the linear operator A and the right-hand side $\mathbf{b}$, we prove the next-level duality theorem, the Farkas lemma for cones. The final step also takes the objective function into account and provides duality theorems for cone programming. The semidefinite case will be dealt with as a corollary.

For people accustomed to the behavior of linear programs, cone programs have some surprises in store. In fact, they seem to misbehave in various ways, as examples in this chapter will illustrate. On a large scale, however, these are only small blunders: Cone programs will turn out to be almost as civilized as linear programs, from which we draw some intuition at various points.

Throughout this chapter we fix real and finite-dimensional vector spaces V and W, equipped with scalar products. In the semidefinite case, we will have $V = \mathrm{SYM}_n$ with the scalar product $\langle X, Y \rangle = X \bullet Y$, and $W = \mathbb{R}^m$ with the standard scalar product $\langle \mathbf{x}, \mathbf{y} \rangle = \mathbf{x}^T \mathbf{y}$.

Actually, the restriction to a finite dimension is not necessary for the material presented in this chapter. With only minimal changes, we could consider V and W as arbitrary Hilbert spaces – we stick to the finite-dimensional setting mainly in the interest of the readers who are not familiar with Hilbert spaces.

4.2 Closed Convex Cones

4.2.1 Definition. *Let $K \subseteq V$ be a nonempty closed set.[1] K is called a closed convex cone if the following two conditions hold.*

(i) *For all $\mathbf{x} \in K$ and all nonnegative real numbers λ, we have $\lambda \mathbf{x} \in K$.*
(ii) *For all $\mathbf{x}, \mathbf{y} \in K$, we have $\mathbf{x} + \mathbf{y} \in K$.*

Condition (i) ensures that K is a cone, while condition (ii) guarantees convexity of K. Indeed, if $\mathbf{x}, \mathbf{y} \in K$ and $\lambda \in [0,1]$, then $(1-\lambda)\mathbf{x}$ and $\lambda \mathbf{y}$ are both in K by (i), and then (ii) shows that $(1-\lambda)\mathbf{x} + \lambda \mathbf{y} \in K$, as required by convexity.

4.2.2 Lemma. *The set $\mathrm{PSD}_n \subseteq \mathrm{SYM}_n$ of positive semidefinite matrices is a closed convex cone.*

Proof. Using the characterization of positive semidefinite matrices provided by Fact 2.2.1(ii), this is easy. If $\mathbf{x}^T M \mathbf{x} \geq 0$ and $\mathbf{x}^T N \mathbf{x} \geq 0$, then also $\mathbf{x}^T \lambda M \mathbf{x} = \lambda \mathbf{x}^T M \mathbf{x} \geq 0$ for $\lambda \geq 0$ and $\mathbf{x}^T (M + N)\mathbf{x} = \mathbf{x}^T M \mathbf{x} + \mathbf{x}^T N \mathbf{x} \geq 0$.

To show closedness, we check that the complement is open. Indeed, if we have a symmetric matrix M that is not positive semidefinite, then there exists $\tilde{\mathbf{x}} \in \mathbb{R}^n$ such that $\mathbf{x}^T M \mathbf{x} < 0$, and this inequality still holds for all matrices M' in a sufficiently small neighborhood of M. $\qquad \square$

Let us look at other examples of closed convex cones. It is obvious that the nonnegative orthant $\mathbb{R}^n_+ = \{\mathbf{x} \in \mathbb{R}^n : \mathbf{x} \geq \mathbf{0}\}$ is a closed convex cone; even more trivial examples of closed convex cones in $\mathbb{R}^n$ are $K = \{\mathbf{0}\}$ and $K = \mathbb{R}^n$. We can also get new cones as direct sums of cones (the proof of the following fact is left to the reader).

4.2.3 Fact. *Let $K \subseteq V$, $L \subseteq W$ be closed convex cones. Then*

$$K \oplus L := \{(\mathbf{x}, \mathbf{y}) \in V \oplus W : \mathbf{x} \in K, \mathbf{y} \in L\}$$

is again a closed convex cone, the direct sum *of K and L.*

Let us recall that $V \oplus W$, the *direct sum* of V and W is the set $V \times W$, turned into a vector space with scalar product via

$$\begin{aligned}
(\mathbf{x}, \mathbf{y}) + (\mathbf{x}', \mathbf{y}') &:= (\mathbf{x} + \mathbf{x}', \mathbf{y} + \mathbf{y}'), \\
\lambda(\mathbf{x}, \mathbf{y}) &:= (\lambda \mathbf{x}, \lambda \mathbf{y}), \\
\langle (\mathbf{x}, \mathbf{y}), (\mathbf{x}', \mathbf{y}') \rangle &:= \langle \mathbf{x}, \mathbf{x}' \rangle + \langle \mathbf{y}, \mathbf{y}' \rangle.
\end{aligned}$$

Now we get to some more interesting cones.

---

[1] We refer to the standard topology on V, induced by the open balls $B(\mathbf{c}, \varrho) = \{\mathbf{x} \in V : \|\mathbf{x} - \mathbf{c}\| < \varrho\}$, where $\|\mathbf{x}\| = \sqrt{\langle \mathbf{x}, \mathbf{x} \rangle}$.

The Ice Cream Cone in $\mathbb{R}^n$

This cone is defined as

$$\nabla_n = \{(\mathbf{x}, r) \in \mathbb{R}^{n-1} \times \mathbb{R} : \|\mathbf{x}\| \leq r\};$$

see Fig. 4.1 for an illustration in $\mathbb{R}^3$ that (hopefully) explains the name. It is closed because of the "$\leq$" (this argument is similar to the one in the proof of Lemma 4.2.2), and its convexity follows from the triangle inequality $\|\mathbf{x} + \mathbf{y}\| \leq \|\mathbf{x}\| + \|\mathbf{y}\|$.

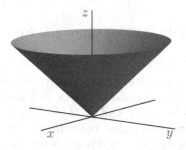

Fig. 4.1 The (lower part of the boundary of the) ice cream cone in $\mathbb{R}^3$

The Toppled Ice Cream Cone in $\mathbb{R}^3$

Here is another closed convex cone, the *toppled ice cream cone*. It is defined as

$$\triangleleft\!0 = \{(x, y, z) \in \mathbb{R}^3 : x \geq 0, y \geq 0, xy \geq z^2\}; \tag{4.3}$$

see Fig. 4.2.

Exercise 4.1 formally explains why we call this the toppled ice cream cone. We remark that $\triangleleft\!0$ can alternatively be defined as the set of all (x, y, z) such that the symmetric matrix

$$\begin{pmatrix} x & z \\ z & y \end{pmatrix}$$

is positive semidefinite. From our earlier considerations, we thus derive

4.2.4 Lemma. $\triangleleft\!0$ *is a closed convex cone.*

It seems that instead of $\triangleleft\!0$, we could equivalently talk about PSD_2, but there is a subtlety here: $\triangleleft\!0$ lives in the vector space $\mathbb{R}^3$, while PSD_2 lives in SYM_2. As a vector space, SYM_2 can be identified with $\mathbb{R}^3$ in an obvious way, but the scalar products are different. Indeed,

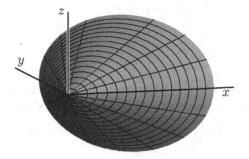

Fig. 4.2 The toppled ice cream cone

$$\begin{pmatrix} x & z \\ z & y \end{pmatrix} \bullet \begin{pmatrix} x' & z' \\ z' & y' \end{pmatrix} = xx' + yy' + 2zz',$$

while

$$(x, y, z)^T (x', y', z') = xx' + yy' + zz'.$$

4.3 Dual Cones

4.3.1 Definition. *Let $K \subseteq V$ be a closed convex cone. The set*

$$K^* := \{ \mathbf{y} \in V : \langle \mathbf{y}, \mathbf{x} \rangle \geq 0 \text{ for all } \mathbf{x} \in K \}$$

is called the dual cone of K.

In fact, K^* is again a closed convex cone. We omit the simple proof; it uses bilinearity of the scalar product to verify the cone conditions, and the Cauchy–Schwarz inequality for closedness.

Let us illustrate this notion on the examples that we have seen earlier. What is the dual of the nonnegative orthant $\mathbb{R}^n_+$? This is the set of all $\mathbf{y}$ such that

$$\mathbf{y}^T \mathbf{x} \geq 0 \text{ for all } \mathbf{x} \geq \mathbf{0}.$$

This set certainly contains the nonnegative orthant $\{ \mathbf{y} \in \mathbb{R}^n : \mathbf{y} \geq \mathbf{0} \}$ itself, but not more: Given $\mathbf{y} \in \mathbb{R}^n$ with $y_i < 0$, we have $\mathbf{y}^T \mathbf{e}_i < 0$, where $\mathbf{e}_i$ is the i-th unit vector (a member of $\mathbb{R}^n_+$), and this proves that $\mathbf{y}$ is not a member of the dual cone $(\mathbb{R}^n_+)^*$. It follows that the dual of $\mathbb{R}^n_+$ is $\mathbb{R}^n_+$: the nonnegative orthant is *self-dual*.

For the "even more trivial" cones, the situation is as follows:

K	K^*
$\{0\}$	$\mathbb{R}^n$
$\mathbb{R}^n$	$\{0\}$

We leave the computation of the dual of the ice cream cone $\bigtriangledown_n$ to the reader (see Exercise 4.2) and proceed with the toppled ice cream cone (4.3).

4.3.2 Lemma. *The dual of the toppled ice cream cone* $\triangleleft$ *is*

$$\triangleleft^* = \{(x,y,z) \in \mathbb{R}^3 : x \geq 0, y \geq 0, xy \geq \frac{z^2}{4}\} \subseteq \mathbb{R}^3,$$

a "vertically stretched" version of $\triangleleft$.

Proof. The computations are not very enlightening, but we want to show at least one nontrivial duality proof. We first want to show the inclusion "$\supseteq$" in the statement of the lemma, and this uses the AGM inequality (arithmetic mean of nonnegative numbers is at least the geometric mean). Let us fix $\tilde{\mathbf{y}} = (\tilde{x}, \tilde{y}, \tilde{z})$ such that $\tilde{x} \geq 0, \tilde{y} \geq 0, \tilde{x}\tilde{y} \geq \tilde{z}^2/4$. For $\mathbf{x} = (x,y,z) \in \triangleleft$ chosen arbitrarily, we get

$$
\begin{aligned}
\tilde{\mathbf{y}}^T \mathbf{x} &= \tilde{x}x + \tilde{y}y + \tilde{z}z \\
&= 2\frac{\tilde{x}x + \tilde{y}y}{2} + \tilde{z}z \\
&\geq 2\sqrt{\tilde{x}x\tilde{y}y} + \tilde{z}z \\
&\geq 2\frac{|\tilde{z}|}{2}|z| + \tilde{z}z \geq 0.
\end{aligned}
$$

This means that $\mathbf{y} \in \triangleleft^*$.

For the other inclusion, let us fix $\tilde{\mathbf{y}} = (\tilde{x}, \tilde{y}, \tilde{z})$ such that $\tilde{x} < 0$ or $\tilde{y} < 0$ or $\tilde{x}\tilde{y} < \tilde{z}^2/4$; we now need to find a proof for $\tilde{\mathbf{y}} \notin \triangleleft^*$. If $\tilde{x} < 0$, we choose $\mathbf{x} = (1,0,0) \in \triangleleft$ and get the desired proof $\tilde{\mathbf{y}}^T\mathbf{x} < 0$. If $\tilde{y} < 0$, $\mathbf{x} = (0,1,0)$ will do. In the case of $\tilde{x}, \tilde{y} \geq 0$ but $\tilde{x}\tilde{y} < \tilde{z}^2/4$, let us first assume that $\tilde{z} \geq 0$ and set

$$\mathbf{x} = (\tilde{y}, \tilde{x}, -\sqrt{\tilde{x}\tilde{y}}) \in \triangleleft.$$

We then compute

$$\tilde{\mathbf{y}}^T\mathbf{x} = 2\tilde{x}\tilde{y} - \tilde{z}\sqrt{\tilde{x}\tilde{y}} < 2\tilde{x}\tilde{y} - 2\tilde{x}\tilde{y} = 0.$$

For $\tilde{z} < 0$, we pick $\mathbf{x} = (\tilde{y}, \tilde{x}, \sqrt{\tilde{x}\tilde{y}}) \in \triangleleft$.   □

We conclude this section with the following intuitive fact: the dual of a direct sum of cones is the direct sum of the dual cones.

4.3.3 Lemma. *Let* $K \subseteq V$, $L \subseteq W$ *be closed convex cones. Then*

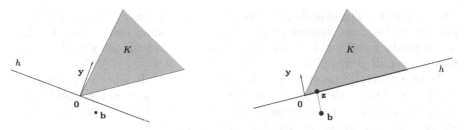

Fig. 4.3 A point **b** not contained in a closed convex cone K can be separated from K by a hyperplane $h = \{\mathbf{x} \in V : \langle \mathbf{y}, \mathbf{x} \rangle = 0\}$ through the origin (*left*). The separating hyperplane resulting from the proof of Theorem 4.4.2 (*right*)

$$(K \oplus L)^* = K^* \oplus L^*.$$

This fact is easy but not entirely trivial. It actually requires a small proof; see Exercise 4.3.

4.4 A Separation Theorem for Closed Convex Cones

Under any meaningful notion of duality, you expect the dual of the dual to be the primal (original) object. For cone duality (Definition 4.3.1), this indeed works.

4.4.1 Lemma. *Let $K \subseteq V$ be a closed convex cone. Then $(K^*)^* = K$.*

Maybe surprisingly, the proof of this innocent-looking fact already requires the machinery of separation theorems which will also be essential for cone programming duality below. Separation theorems generally assert that disjoint convex sets can be separated by a hyperplane.

The following is arguably the simplest nontrivial separation theorem. On top of elementary calculations, the proof only requires one standard result from analysis.

4.4.2 Theorem. *Let $K \subseteq V$ be a closed convex cone, and let $\mathbf{b} \in V \setminus K$. Then there exists a vector $\mathbf{y} \in V$ such that*

$$\langle \mathbf{y}, \mathbf{x} \rangle \geq 0 \text{ for all } \mathbf{x} \in K, \text{ and } \langle \mathbf{y}, \mathbf{b} \rangle < 0.$$

The statement is illustrated in Fig. 4.3 (left). We also say that $\mathbf{y}$ is a *witness* for $\mathbf{b} \notin K$.

Proof. The plan of the proof is straightforward: We let $\mathbf{z}$ be the point of K nearest to $\mathbf{b}$ (in the distance $\|\mathbf{z} - \mathbf{b}\| = \sqrt{\langle \mathbf{z} - \mathbf{b}, \mathbf{z} - \mathbf{b} \rangle}$ induced by the scalar product), and we check that the vector $\mathbf{y} = \mathbf{z} - \mathbf{b}$ is as required; see

Fig. 4.3 (right). The existence of $\mathbf{z}$ follows from the general theory: If C is a nonempty closed and convex set in a finite-dimensional vector space V with scalar product and $\mathbf{x} \in V$ is arbitrary, then there is a unique $\mathbf{y} \in C$ nearest to $\mathbf{x}$ among all points of C; see [EMT04, Sect. 2.2b].

With $\mathbf{z}$ a nearest point of K to $\mathbf{b}$, we set $\mathbf{y} = \mathbf{z} - \mathbf{b}$. First we check that $\langle \mathbf{y}, \mathbf{z} \rangle = 0$. This is clear for $\mathbf{z} = \mathbf{0}$. For $\mathbf{z} \neq \mathbf{0}$, if $\mathbf{z}$ were not perpendicular to $\mathbf{y}$, we could move $\mathbf{z}$ slightly along the ray $\{t\mathbf{z} : t \geq 0\} \subseteq K$ and get a point closer to $\mathbf{b}$ (here we use the fact that K is a cone).

More formally, let us first assume that $\langle \mathbf{y}, \mathbf{z} \rangle > 0$, and let us set $\mathbf{z}' = (1 - \alpha)\mathbf{z}$ for a small $\alpha > 0$. We calculate $\|\mathbf{z}' - \mathbf{b}\|^2 = \langle (\mathbf{y} - \alpha\mathbf{z}), (\mathbf{y} - \alpha\mathbf{z}) \rangle = \|\mathbf{y}\|^2 - 2\alpha\langle \mathbf{y}, \mathbf{z} \rangle + \alpha^2\|\mathbf{z}\|^2$. We have $2\alpha\langle \mathbf{y}, \mathbf{z} \rangle > \alpha^2\|\mathbf{z}\|^2$ for all sufficiently small $\alpha > 0$, and thus $\|\mathbf{z}' - \mathbf{b}\|^2 < \|\mathbf{y}\|^2 = \|\mathbf{z} - \mathbf{b}\|^2$. This contradicts $\mathbf{z}$ being a nearest point. The case $\langle \mathbf{y}, \mathbf{z} \rangle < 0$ is handled through $\mathbf{z}' = (1 + \alpha)\mathbf{z}$.

To verify $\langle \mathbf{y}, \mathbf{b} \rangle < 0$, we recall that $\mathbf{y} \neq \mathbf{0}$, and we compute $0 < \langle \mathbf{y}, \mathbf{y} \rangle = \langle \mathbf{y}, \mathbf{z} \rangle - \langle \mathbf{y}, \mathbf{b} \rangle = -\langle \mathbf{y}, \mathbf{b} \rangle$.

Next, let $\mathbf{x} \in K$, $\mathbf{x} \neq \mathbf{z}$. The angle $\angle\mathbf{bzx}$ has to be at least $90°$, for otherwise, points on the segment $\mathbf{zx}$ sufficiently close to $\mathbf{z}$ would lie closer to $\mathbf{b}$ than $\mathbf{z}$ (here we use convexity of K); equivalently, $\langle (\mathbf{b}-\mathbf{z}), (\mathbf{x}-\mathbf{z}) \rangle \leq 0$. This is similar to the above argument for $\langle \mathbf{y}, \mathbf{z} \rangle = 0$ and we leave a formal verification to the reader. Thus $0 \geq \langle (\mathbf{b} - \mathbf{z}), (\mathbf{x} - \mathbf{z}) \rangle = -\langle \mathbf{y}, \mathbf{x} \rangle + \langle \mathbf{y}, \mathbf{z} \rangle = -\langle \mathbf{y}, \mathbf{x} \rangle$. $\square$

Using this result, we can now show that $(K^*)^* = K$ for every closed convex cone.

Proof of Lemma 4.4.1. For the direction $K \subseteq (K^*)^*$, we just need to apply the definition of duality: Let us choose $\mathbf{b} \in K$. By definition of K^*, $\langle \mathbf{y}, \mathbf{b} \rangle = \langle \mathbf{b}, \mathbf{y} \rangle \geq 0$ for all $\mathbf{y} \in K^*$; this shows that $\mathbf{b} \in (K^*)^*$.

For the other direction, we let $\mathbf{b} \in V \setminus K$. According to Theorem 4.4.2, we find a vector $\mathbf{y}$ such $\langle \mathbf{y}, \mathbf{x} \rangle \geq 0$ for all $\mathbf{x} \in K$ and $\langle \mathbf{y}, \mathbf{b} \rangle = \langle \mathbf{b}, \mathbf{y} \rangle < 0$. The former inequality shows that $\mathbf{y} \in K^*$, but then the latter inequality shows that $\mathbf{b} \notin (K^*)^*$. $\square$

4.5 The Farkas Lemma, Cone Version

The Farkas lemma is a cornerstone of linear programming theory. It comes in several equivalent versions, among them the following.

4.5.1 Lemma (Farkas). *Let $A \in \mathbb{R}^{m \times n}$ be an $m \times n$ matrix, and let $\mathbf{b} \in \mathbb{R}^m$. Then*

- *Either the system $A\mathbf{x} = \mathbf{b}$, $\mathbf{x} \geq \mathbf{0}$ has a solution $\mathbf{x} \in \mathbb{R}^n$.*
- *Or the system $A^T\mathbf{y} \geq \mathbf{0}$, $\mathbf{b}^T\mathbf{y} < 0$ has a solution $\mathbf{y} \in \mathbb{R}^m$.*

but not both.

There is also a more general version of the Farkas lemma, which yields a *certificate* for the unsolvability of an arbitrary unsolvable system of linear equations and inequalities.

Here, we first want to make the point that the Farkas lemma is a special case of Theorem 4.4.2. For this, we define $V = \mathbb{R}^m$ and

$$K = \{A\mathbf{x} : \mathbf{x} \in \mathbb{R}^n_+\} \subseteq V.$$

Thus, K consists of all nonnegative linear combinations of columns of A. We call this a *finitely generated cone*. Finitely generated cones are closed and convex, and hence Theorem 4.4.2 applies.

Now the first system $A\mathbf{x} = \mathbf{b}$, $\mathbf{x} \geq \mathbf{0}$ having no solution just means that $\mathbf{b} \in V \setminus K$. By Theorem 4.4.2, there exists $\mathbf{y} \in V = \mathbb{R}^m$ such that

$$\mathbf{y}^T A\mathbf{x} \geq 0 \text{ for all } \mathbf{x} \in \mathbb{R}^n_+, \text{and } \mathbf{y}^T \mathbf{b} < 0.$$

Since the former inequality means $A^T\mathbf{y} \in (\mathbb{R}^n_+)^* = \mathbb{R}^n_+$, we indeed get a solution to the second system $A^T\mathbf{y} \geq \mathbf{0}$, $\mathbf{b}^T\mathbf{y} < 0$. It remains to observe that the first system and the second one can never be solvable simultaneously. (To see this, we premultiply the first system with a transposed solution of the second one).

In this section, we want to generalize the Farkas lemma to deal with systems of the form

$$A(\mathbf{x}) = \mathbf{b}, \mathbf{x} \in K,$$

where $K \subseteq V$ is some closed convex cone, and A is a linear operator from V to W.

The "standard" Farkas lemma deals with the case $K = \mathbb{R}^n_+ \subseteq V := \mathbb{R}^n$, $W = \mathbb{R}^m$, where a linear operator can be represented by a matrix. For semidefinite programming, on the other hand, we need to consider the case $K = \text{PSD}_n \subseteq V := \text{SYM}_n$ and $W = \mathbb{R}^m$.

There are two obstacles to overcome, the first one being merely technical: we need to define what A^T is supposed to mean for a general linear operator. The second obstacle is the real one: a cone of the form $\{A(\mathbf{x}) : \mathbf{x} \in K\}$ is convex, but not necessarily closed (we will see an example), so that Theorem 4.4.2 may not be applicable. We now address both obstacles in turn.

Adjoint Operators

Here is the appropriate generalization of the transposed matrix.

4.5.2 Definition. *Let $A: V \rightarrow W$ be a linear operator. A linear operator $A^T: W \rightarrow V$ is called an adjoint of A if*

$$\langle \mathbf{y}, A(\mathbf{x}) \rangle = \langle A^T(\mathbf{y}), \mathbf{x} \rangle \text{ for all } \mathbf{x} \in V \text{ and } \mathbf{y} \in W.$$

For $V = \mathbb{R}^n$, $W = \mathbb{R}^m$ and A represented by an $m \times n$ matrix, the transposed matrix represents the unique adjoint of A. More generally, if V and W are finite-dimensional (which we assume), there is an adjoint A^T of A. And *if* there is an adjoint, then it is easy to see that it is unique, which justifies the notation A^T.

In semidefinite programming, we have $V = \mathrm{SYM}_n$ (the scalar product[2] is $X \bullet Y$), and $W = \mathbb{R}^m$ (with standard scalar product). Also in this case the adjoint is easy to determine.

4.5.3 Lemma. *Let $V = \mathrm{SYM}_n, W = \mathbb{R}^m$, and $A: V \to W$ defined by $A(X) = (A_1 \bullet X, A_2 \bullet X, \ldots, A_m \bullet X)$. Then*

$$A^T(\mathbf{y}) = \sum_{i=1}^m y_i A_i.$$

Proof. We compute

$$\langle \mathbf{y}, A(X) \rangle \; := \; \mathbf{y}^T A(X)$$
$$= \sum_{i=1}^m y_i (A_i \bullet X) = (\sum_{i=1}^m y_i A_i) \bullet X = A^T(\mathbf{y}) \bullet X =: \langle A^T(\mathbf{y}), X \rangle.$$

Here we have used the linearity of $\bullet$ in the first argument.   □

In order to stay as close as possible to the familiar matrix terminology, we will also introduce the following notation. If $V_1, V_2, \ldots, V_n$ and $W_1, W_2, \ldots, W_m$ are vector spaces with scalar products, and if $A_{ij}: V_j \to W_i$ are linear operators for all i, j, then we write the "matrix"

$$\begin{pmatrix} A_{11} & A_{12} & \cdots & A_{1n} \\ A_{21} & A_{22} & \cdots & A_{2n} \\ \vdots & \vdots & & \vdots \\ A_{m1} & A_{m2} & \cdots & A_{mn} \end{pmatrix} \tag{4.4}$$

for the linear operator $\bar{A}: V_1 \oplus V_2 \oplus \cdots \oplus V_n \to W_1 \oplus W_2 \oplus \cdots \oplus W_m$ defined by

$$\bar{A}(\mathbf{x}_1, \mathbf{x}_2, \ldots, \mathbf{x}_n) = \left(\sum_{j=1}^n A_{1j}(\mathbf{x}_j), \sum_{j=1}^n A_{2j}(\mathbf{x}_j), \ldots, \sum_{j=1}^n A_{mj}(\mathbf{x}_j) \right).$$

A simple calculation then shows that

---

[2] At this point, a skeptical reader might want to convince him- or herself that this is indeed a scalar product.

$$\bar{A}^T = \left(\begin{array}{c|c|c|c} A_{11}^T & A_{21}^T & \cdots & A_{m1}^T \\ \hline A_{12}^T & A_{22}^T & \cdots & A_{m2}^T \\ \hline \vdots & \vdots & & \vdots \\ \hline A_{1n}^T & A_{2n}^T & \cdots & A_{mn}^T \end{array} \right), \tag{4.5}$$

just as with matrices.

For the remainder of this chapter, we fix a linear operator A from V to W.

The Farkas Lemma, Bogus Version

We would like to be able to claim the following.

Let $K \subseteq V$ be a closed convex cone, and let $\mathbf{b} \in W$. *Either* the system

$$A(\mathbf{x}) = \mathbf{b}, \mathbf{x} \in K \tag{4.6}$$

has a solution $\mathbf{x} \in V$, *or* the system

$$A^T(\mathbf{y}) \in K^*, \langle \mathbf{b}, \mathbf{y} \rangle < 0 \tag{4.7}$$

has a solution, but not both.

Indeed, this follows from Theorem 4.4.2 along the lines of what we did above for the standard Farkas lemma, *provided that* the cone $C := A(K) = \{A(\mathbf{x}) : \mathbf{x} \in K\}$ is closed. But this need not be true. Here is an example.

Let $K = \triangleleft\!\!\!\bigcirc$, the toppled ice cream cone (4.3) in $\mathbb{R}^3$, and let

$$A = \begin{pmatrix} 0 & 1 & 0 \\ 0 & 0 & 1 \end{pmatrix}.$$

Then C is the projection of $\triangleleft\!\!\!\bigcirc$ onto the yz-plane – see Fig. 4.2 on page 49 for the geometric intuition – and assumes the form

$$C = \{(y, z) \in \mathbb{R}^2 : (x, y, z) \in \triangleleft\!\!\!\bigcirc\} = \mathbf{0} \cup (\{y \in \mathbb{R} : y > 0\} \times \mathbb{R}),$$

a set that is obviously not closed.

In such cases, Theorem 4.4.2 is not applicable, and the Farkas lemma as envisioned above may fail. Such a failure can already be constructed from our previous example. If we in addition set $\mathbf{b} = (0, 1) \in \mathbb{R}^2$, *both* (4.6) and (4.7) are unsolvable (we encourage the reader to go through this example in detail).

The Farkas Lemma, Cone Version

To save the situation, we work with the closure of C.

4.5.4 Lemma. *Let $K \subseteq V$ be a closed convex cone, and $C = \{A(\mathbf{x}) : \mathbf{x} \in K\}$. Then $\overline{C}$, the closure of C, is a closed convex cone.*

Proof. The closure of C is the set of all limit points of C. Formally, $\mathbf{b} \in \overline{C}$ if and only if there exists a sequence $(\mathbf{y}_k)_{k \in \mathbb{N}}$ such that $\mathbf{y}_k \in C$ for all k and $\lim_{k \to \infty} \mathbf{y}_k = \mathbf{b}$. This yields that $\overline{C}$ is a convex cone, using that C is a convex cone. In addition, $\overline{C}$ is closed. $\square$

The fact "$\mathbf{b} \in \overline{C}$" can be formulated without reference to the cone C, which will be more convenient in what follows.

4.5.5 Definition. *Let $K \subseteq V$ be a closed convex cone. The system*

$$A(\mathbf{x}) = \mathbf{b}, \ \mathbf{x} \in K$$

is called limit-feasible if there exists a sequence $(\mathbf{x}_k)_{k \in \mathbb{N}}$ such that $\mathbf{x}_k \in K$ for all $k \in \mathbb{N}$ and

$$\lim_{k \to \infty} A(\mathbf{x}_k) = \mathbf{b}.$$

It is clear that if $A(\mathbf{x}) = \mathbf{b}, \ \mathbf{x} \in K$ is limit-feasible, then $\mathbf{b} \in \overline{C}$, but the reverse implication also holds. If $(\mathbf{y}_k)_{k \in \mathbb{N}}$ is a sequence in C converging to $\mathbf{b}$, then any sequence $(\mathbf{x}_k)_{k \in \mathbb{N}}$ such that $\mathbf{y}_k = A(\mathbf{x}_k)$ for all k proves the limit-feasibility of our system. Here is the correct Farkas lemma for equation systems over closed convex cones.

4.5.6 Lemma (Farkas lemma for cones). *Let $K \subseteq V$ be a closed convex cone, and $\mathbf{b} \in W$. Either the system*

$$A(\mathbf{x}) = \mathbf{b}, \mathbf{x} \in K$$

is limit-feasible, or the system

$$A^T(\mathbf{y}) \in K^*, \langle \mathbf{b}, \mathbf{y} \rangle < 0$$

has a solution, but not both.

Proof. If $A(\mathbf{x}) = \mathbf{b}, \ \mathbf{x} \in K$ is limit-feasible, we choose any sequence $(\mathbf{x}_k)_{k \in \mathbb{N}}$ that proves its limit feasibility. For $\mathbf{y} \in W$, we compute

$$\langle \mathbf{y}, \mathbf{b} \rangle = \langle \mathbf{y}, \lim_{k \to \infty} A(\mathbf{x}_k) \rangle = \lim_{k \to \infty} \langle \mathbf{y}, A(\mathbf{x}_k) \rangle = \lim_{k \to \infty} \langle A^T(\mathbf{y}), \mathbf{x}_k \rangle.$$

If $A^T(\mathbf{y}) \in K^*$, then $\mathbf{x}_k \in K$ yields $\langle A^T(\mathbf{y}), \mathbf{x}_k \rangle \geq 0$ for all $k \in \mathbb{N}$, and $\langle \mathbf{y}, \mathbf{b} \rangle \geq 0$ follows. Thus, the second system has no solution.

If $A(\mathbf{x}) = \mathbf{b}$, $\mathbf{x} \in K$ is not limit-feasible, this can be expressed equivalently as $\mathbf{b} \notin \overline{C}$, where $C = \{A(\mathbf{x}) : \mathbf{x} \in K\}$. Since $\overline{C}$ is a closed convex cone, we can apply Theorem 4.4.2 and obtain a hyperplane that strictly separates $\mathbf{b}$ from $\overline{C}$ (and in particular from C). This means that we find $\mathbf{y} \in W$ such that

$$\langle \mathbf{y}, \mathbf{b} \rangle < 0 \text{ and for all } \mathbf{x} \in K, \ \langle \mathbf{y}, A(\mathbf{x}) \rangle = \langle A^T(\mathbf{y}), \mathbf{x} \rangle \geq 0.$$

It remains to observe that the statement "$\langle A^T(\mathbf{y}), \mathbf{x} \rangle \geq 0$ for all $\mathbf{x} \in K$" is equivalent to "$A^T(\mathbf{y}) \in K^*$."   □

4.6 Cone Programs

Here is the definition of a cone program, in a form somewhat more general than we would need for semidefinite programming. This extra generality will introduce symmetry between the primal and the dual program.

4.6.1 Definition. *Let $K \subseteq V$, $L \subseteq W$ be closed convex cones, let $\mathbf{b} \in W$, $\mathbf{c} \in V$, and let $A: V \to W$ be a linear operator. A* cone program *is an optimization problem of the form*

$$\begin{aligned}
\text{Maximize} \quad & \langle \mathbf{c}, \mathbf{x} \rangle \\
\text{subject to} \quad & \mathbf{b} - A(\mathbf{x}) \in L \\
& \mathbf{x} \in K.
\end{aligned} \qquad (4.8)$$

For $L = \{\mathbf{0}\}$, we get cone programs in *equational form*.

Following the linear programming case, we call the cone program *feasible* if there is some *feasible solution*, i.e., a vector $\tilde{\mathbf{x}}$ with $\mathbf{b} - A(\tilde{\mathbf{x}}) \in L$, $\tilde{\mathbf{x}} \in K$. The *value* of a feasible cone program is defined as

$$\sup\{\langle \mathbf{c}, \mathbf{x} \rangle : \mathbf{b} - A(\mathbf{x}) \in L, \mathbf{x} \in K\}, \qquad (4.9)$$

which includes the possibility that the value is ∞.

An *optimal solution* is a feasible solution $\mathbf{x}^*$ such that $\langle \mathbf{c}, \mathbf{x}^* \rangle \geq \langle \mathbf{c}, \mathbf{x} \rangle$ for all feasible solutions $\mathbf{x}$. Consequently, if there is an optimal solution, then the value of the cone program is finite, and that value is attained, meaning that the supremum in (4.9) is a maximum.

Reachability of the Value

There are cone programs with finite value but no optimal solution. Here is an example involving the toppled ice cream cone; see page 48 (we leave it to the reader to put it into the form (4.8) via suitable $A, \mathbf{b}, \mathbf{c}, K, L$):

$$\text{Minimize} \quad x$$
$$\text{subject to} \quad z = 1 \qquad\qquad (4.10)$$
$$(x, y, z) \in \triangleleft.$$

After substituting $z = 1$ into the definition of the toppled ice cream cone, we obtain the following equivalent constrained optimization problem in two variables:

$$\text{Minimize} \quad x$$
$$\text{subject to} \quad x \geq 0$$
$$xy \geq 1.$$

It is clear that the value of x is bounded from below by 0, and that values arbitrarily close to 0 can be attained. Due to the constraint $xy \geq 1$, however, the value 0 itself cannot be attained. This means that the cone program (4.10) has value 0 but no optimal solution.

The Limit Value

Another aspect that we do not see in linear programming is that of limit feasibility, a notion that we already introduced for equation systems over cones – see Definition 4.5.5. If a linear program is infeasible, then it will remain infeasible under any sufficiently small perturbation of the right-hand side $\mathbf{b}$. In contrast, there are infeasible cone programs that become feasible under an *arbitrarily* small perturbation of $\mathbf{b}$.

The following is merely a repetition of Definition 4.5.5 for the linear operator $(A \mid \text{id}) \colon V \oplus W \to W$ (recall the matrix notation introduced on page 54) and the cone $K \oplus L$.

4.6.2 Definition. *The cone program (4.8) is called limit-feasible if there exist sequences $(\mathbf{x}_k)_{k \in \mathbb{N}}$ and $(\mathbf{x}'_k)_{k \in \mathbb{N}}$ such that $\mathbf{x}_k \in K$ and $\mathbf{x}'_k \in L$ for all $k \in N$, and*

$$\lim_{k \to \infty} (A(\mathbf{x}_k) + \mathbf{x}'_k) = \mathbf{b}.$$

Such sequences $(\mathbf{x}_k)_{k \in \mathbb{N}}$ and $(\mathbf{x}'_k)_{k \in \mathbb{N}}$ are called feasible sequences of (4.8).

We stress that the feasible sequences themselves are not required to converge (see also the example on page 59 below).

Every feasible program is limit-feasible, but the converse is guaranteed only if the cone $C = \{A(\mathbf{x}) + \mathbf{x}' : \mathbf{x} \in K, \mathbf{x}' \in L\}$ is closed. For $L = \{\mathbf{0}\}$, we have seen a counterexample on page 55.

We can assign a value even to a limit-feasible cone program, and we call it the limit value.

4.6.3 Definition. *Given a feasible sequence $(\mathbf{x}_k)_{k \in \mathbb{N}}$ of a cone program (4.8), we define its value as*

$$\langle \mathbf{c}, (\mathbf{x}_k)_{k\in\mathbb{N}}\rangle := \limsup_{k\to\infty}\langle \mathbf{c}, \mathbf{x}_k\rangle.$$

The *limit value* of (4.8) is then defined as

$$\sup\{\langle \mathbf{c}, (\mathbf{x}_k)_{k\in\mathbb{N}}\rangle : (\mathbf{x}_k)_{k\in\mathbb{N}} \text{ is a feasible sequence of (4.8)}\}.$$

It is not hard to check (Exercise 4.6) that the limit value is actually attained by some feasible sequence.

Value vs. Limit Value

By definition, the value of a feasible cone program is always bounded above by its limit value, and it is tempting to think that the two are equal. But this is not true in general. There are feasible programs with finite value but infinite limit value. Here is one such program in two variables x and z:

$$\text{Maximize} \quad z$$
$$\text{subject to} \quad (x, 0, z) \in \triangleleft.$$

The program is feasible (choose $x \geq 0$, $z = 0$). Every feasible solution must satisfy $z = 0$, so the value of the program is 0. For the following computation of the limit value, let us explicitly write down the parameters of this program in the form of (4.8):

$$\mathbf{c} = (0, 1), \ A = \begin{pmatrix} -1 & 0 \\ 0 & 0 \\ 0 & -1 \end{pmatrix}, \ \mathbf{b} = (0, 0, 0), \ L = \triangleleft, \ K = \mathbb{R}^2.$$

Now let us define

$$\mathbf{x}_k = (k^3, k) \in \mathbb{R}^2, \ \mathbf{x}'_k = (k^3, 1/k, k) \in \triangleleft, \quad k \in \mathbb{N}.$$

Then we get

$$\lim_{k\to\infty} A\mathbf{x}_k + \mathbf{x}'_k = \lim_{k\to\infty} \left((-k^3, 0, -k) + (k^3, 1/k, k)\right) = \lim_{k\to\infty} (0, 1/k, 0) = \mathbf{0} = \mathbf{b}.$$

This means that $(\mathbf{x}_k)_{k\in\mathbb{N}}$ and $(\mathbf{x}'_k)_{k\in\mathbb{N}}$ are feasible sequences. The program's limit value is therefore at least

$$\limsup_{k\to\infty} \mathbf{c}^T \mathbf{x}_k = \limsup_{k\to\infty} k = \infty,$$

different from the value.

We may also have a gap if both value and limit value are finite. Adding the constraint $z \leq 1$ to the previous program (which can be done by changing A, $\mathbf{b}$, L accordingly) results in limit value 1, while the value stays at 0.

Fortunately, such pathologies disappear if the program has an *interior point*. In general, requiring additional conditions, with the goal of avoiding exceptional situations, is called *constraint qualification*. Requiring an interior point is known as *Slater's constraint qualification*.

4.6.4 Definition. *An interior point (or Slater point) of the cone program (4.8) is a point* $\mathbf{x}$ *such that*

$$\mathbf{x} \in K, \quad \mathbf{b} - A(\mathbf{x}) \in L,$$

and the following additional requirement holds:

$$\mathbf{x} \in \text{int}(K) \quad \text{if } L = \{\mathbf{0}\}, \text{ and}$$
$$\mathbf{b} - A(\mathbf{x}) \in \text{int}(L) \quad \text{otherwise.}$$

(For a set S, $\text{int}(S)$ is the set of all points $\mathbf{x} \in S$ such that a sufficiently small ball around $\mathbf{x}$ is fully contained in S.)

The case $L = \{\mathbf{0}\}$ (equational form) receives a special treatment in this definition; if L is some other cone that is not full-dimensional, Slater's constraint qualification does not apply. Now we can prove the following.

4.6.5 Theorem. *If the cone program (4.8) has an interior point (which, in particular, means that it is feasible), then the value equals the limit value.*

Proof. Let γ' be the limit value, and let $\varepsilon > 0$ be a real number. The strategy is to construct a feasible solution $\overline{\mathbf{x}}$ such that

$$\langle \mathbf{c}, \overline{\mathbf{x}} \rangle \geq \gamma' - \varepsilon. \tag{4.11}$$

This proves the theorem since it implies that there are feasible solutions of value arbitrarily close to the limit value. The construction is somewhat different for $L = \{\mathbf{0}\}$ and the other cases, but the approach is the same.

We choose feasible sequences $(\mathbf{x}_k)_{k \in \mathbb{N}}$, $(\mathbf{x}'_k)_{k \in \mathbb{N}}$ that attain the program's limit value γ' (possibly ∞). Then we slightly modify the $\mathbf{x}_k$ such that we obtain a sequence $(\mathbf{w}_k)_{k \in \mathbb{N}}$ satisfying

$$\mathbf{w}_k \in K, \quad \mathbf{b} - A(\mathbf{w}_k) \in L, \quad \|\mathbf{x}_k - \mathbf{w}_k\| \leq \delta \tag{4.12}$$

for almost all k, where δ is chosen such that $|\langle \mathbf{c}, \mathbf{x}_k \rangle - \langle \mathbf{c}, \mathbf{w}_k \rangle| < \varepsilon$. Once we have this, we get

$$\limsup_{k \to \infty} \langle \mathbf{c}, \mathbf{w}_k \rangle > \limsup_{k \to \infty} \langle \mathbf{c}, \mathbf{x}_k \rangle - \varepsilon = \gamma' - \varepsilon,$$

meaning that for some value of k, $\bar{\mathbf{x}} = \mathbf{w}_k$ is a feasible solution that satisfies (4.11).

It remains to transform the $\mathbf{x}_k$ into *feasible* $\mathbf{w}_k$, and here the interior point provides the necessary room. Let us fix such an interior point $\tilde{\mathbf{x}}$.

The case $L \neq \{0\}$. With $\xi > 0$ a real number, we define

$$\mathbf{w}_k := (1 - \xi)\mathbf{x}_k + \xi\tilde{\mathbf{x}}.$$

We can guarantee $\|\mathbf{x}_k - \mathbf{w}_k\| \leq \delta$ by choosing ξ small enough. It is also clear that $\mathbf{w}_k \in K$. To establish (4.12), it remains to show that $\mathbf{b} - A(\mathbf{w}_k) \in L$ for almost all k. For this, we use the definition of feasible sequences and compute

$$\mathbf{b} - A(\mathbf{w}_k) = (1 - \xi)(\underbrace{\mathbf{b} - A(\mathbf{x}_k) - \mathbf{x}_k'}_{\to 0} + \underbrace{\mathbf{x}_k'}_{\in L}) + \xi\underbrace{(\mathbf{b} - A(\tilde{\mathbf{x}}))}_{\in \mathrm{int}(L)}.$$

We also have $\xi(\mathbf{b} - A(\tilde{\mathbf{x}})) \in \mathrm{int}(L)$, with a small ball of radius ϱ around it fully contained in L. Adding $(1 - \xi)\mathbf{x}_k' \in L$ takes us to another point $\mathbf{p}_k \in L$ to which the same ϱ applies. We thus have

$$\mathbf{b} - A(\mathbf{w}_k) = \mathbf{p}_k + (1 - \xi)(\mathbf{b} - A(\mathbf{x}_k) - \mathbf{x}_k'),$$

where $\|(1-\xi)(\mathbf{b} - A(\mathbf{x}_k) - \mathbf{x}_k')\| < \varrho$ for k large enough, hence $\mathbf{b} - A(\mathbf{w}_k) \in L$.

The case $L = \{0\}$. Recalling that our vector spaces $V \supseteq K$ and $W \supseteq L$ are finite-dimensional, we choose an arbitrary basis $\mathbf{v}_1, \ldots, \mathbf{v}_n$ of V. The image of A denoted by $\mathrm{Im}(A)$ is spanned by the vectors $A(\mathbf{v}_1), \ldots, A(\mathbf{v}_n)$, and we choose a subset of vectors that form a basis of $\mathrm{Im}(A)$. W.l.o.g. let these be $A(\mathbf{v}_1), \ldots, A(\mathbf{v}_m)$, and let $V' \subseteq V$ be the span of $\mathbf{v}_1, \ldots, \mathbf{v}_m$. It follows that the restriction of A to V' is a bijection between V' and the image of A. Let $A^{-1} \colon \mathrm{Im}(A) \to V'$ be the inverse mapping. It is also a linear mapping between finite-dimensional vector spaces, and as such it is continuous. With $\xi > 0$, we define

$$\mathbf{w}_k := (1 - \xi)(\mathbf{x}_k + A^{-1}(\underbrace{\mathbf{b} - A(\mathbf{x}_k)}_{\in \mathrm{Im}(A)})) + \xi\tilde{\mathbf{x}},$$

where $\tilde{\mathbf{x}}$ is our interior point ($\mathbf{b} \in \mathrm{Im}(A)$ follows from the feasibility of $\tilde{\mathbf{x}}$).

By the construction we have $A(\mathbf{w}_k) = \mathbf{b}$. Since $\lim_{k \to \infty} A(\mathbf{x}_k) = \mathbf{b}$, we also have $\lim_{k \to \infty} A^{-1}(\mathbf{b} - A(\mathbf{x}_k)) = \mathbf{0}$, so we can guarantee $\|\mathbf{x}_k - \mathbf{w}_k\| \leq \delta$ for almost all k, by choosing ξ small enough.

In order to get (4.12), it thus remains to show that $\mathbf{w}_k \in K$ for sufficiently large k. We have $\xi\tilde{\mathbf{x}} \in \mathrm{int}(K)$, with a small ball of some radius ϱ around it fully contained in K. Adding $(1-\xi)\mathbf{x}_k \in K$ yields another point $\mathbf{p}_k \in \mathrm{int}(K)$ to which the same ϱ applies. We thus have

$$\mathbf{w}_k = \mathbf{p}_k + (1 - \xi)A^{-1}(\mathbf{b} - A(\mathbf{x}_k)),$$

where $\|(1 - \xi)A^{-1}(\mathbf{b} - A(\mathbf{x}_k))\| < \varrho$ for k sufficiently large, and so $\mathbf{w}_k \in K$.

$\square$

4.7 Duality of Cone Programming

For this section, let us call the cone program (4.8) the *primal program* and denote it by (P):

$$\text{(P)} \quad \begin{array}{ll} \text{Maximize} & \langle \mathbf{c}, \mathbf{x} \rangle \\ \text{subject to} & \mathbf{b} - A(\mathbf{x}) \in L \\ & \mathbf{x} \in K. \end{array}$$

Then we define its *dual program* as the cone program

$$\text{(D)} \quad \begin{array}{ll} \text{Minimize} & \langle \mathbf{b}, \mathbf{y} \rangle \\ \text{subject to} & A^T(\mathbf{y}) - \mathbf{c} \in K^* \\ & \mathbf{y} \in L^*. \end{array}$$

Formally, this does not have the cone program format (4.8), but we could easily achieve this if necessary by rewriting (D) as follows.

$$\text{(D')} \quad \begin{array}{ll} \text{Maximize} & -\langle \mathbf{b}, \mathbf{y} \rangle \\ \text{subject to} & -\mathbf{c} + A^T(\mathbf{y}) \in K^* \\ & \mathbf{y} \in L^*. \end{array}$$

Having done this, we can also compute the dual of (D') which takes us, not surprisingly, back to (P).

For the dual program (D), which is a minimization problem, the value and limit value are defined through inf's and lim inf's in the canonical way.

Similar to linear programming, we assume that the primal program (P) is feasible and has a finite value. Then we want to conclude that the dual program (P) is also feasible and has the same value. But, unlike in linear programming, we need another condition to make this work: (P) needs to have an interior point. Here is the *Strong Duality Theorem of Cone Programming*.

4.7.1 Theorem. *If the primal program (P) is feasible, has a finite value γ and has an interior point $\tilde{\mathbf{x}}$, then the dual program (D) is also feasible and has the same value γ.*

Here is an outline of the proof. First, we prove *weak duality*: If the primal program (P) is limit-feasible, the limit value of (P) is bounded above by the value of (D), given that (D) is feasible. The proof of this is a no-brainer and follows more or less directly from the definitions of the primal and the dual. Still, weak duality has the important consequence that a feasible (D) – a minimization problem – has a finite value if (P) is limit-feasible.

Then we prove *regular duality*: if (P) is limit-feasible, then the dual program (D) is actually feasible, and there is no gap between the limit value of (P) and the value of (D). For linear programming where there is no difference between value and limit value (see Exercise 4.7), we would be done at this point and would have proved the strong duality theorem. But here the following scenario is still possible: both (P) and (D) are feasible, but there is a gap between their values γ and β; see Fig. 4.4. We can indeed derive an example for this scenario from the cone program with value 0 and limit value 1 that we have constructed on page 59.

Fig. 4.4 Gap between the values of the primal and dual cone programs

To get strong duality, we apply Slater's constraint qualification: If the primal program has an interior point (Definition 4.6.4), then there is no gap between primal and dual value. This result is a trivial consequence of regular duality together with Theorem 4.6.5 (the value equals the limit value in the case where there is an interior point).

The presentation essentially follows Duffin's original article [Duf56] in the classic book *Linear Inequalities and Related Systems*, 1956. This book, edited by Kuhn and Tucker, contains articles by many of the "grand old men," including Dantzig, Ford, Fulkerson, Gale, and Kruskal.

Weak Duality

Let us start with the *weak Duality Theorem*.

4.7.2 Theorem. *If the dual program (D) is feasible, and if the primal program (P) is limit-feasible, then the limit value of (P) is bounded above by the value of (D).*

If (P) is feasible as well, this implies that the value of (P) is bounded from above by the value of (D), and that both values are finite. This is weak duality as we know it from linear programming.

Proof. We pick any feasible solution $\mathbf{y}$ of (D) and arbitrary feasible sequences $(\mathbf{x}_k)_{k \in \mathbb{N}}$, $(\mathbf{x}'_k)_{k \in \mathbb{N}}$ of (P). Using the defining property of the adjoint (Definition 4.5.2), we get

$$0 \leq \underbrace{\langle A^T(\mathbf{y}) - \mathbf{c},\ \mathbf{x}_k\ \rangle}_{\in K^* \quad \in K} + \langle\ \underbrace{\mathbf{y}}_{\in L^*}\ ,\ \underbrace{\mathbf{x}'_k}_{\in L}\ \rangle = \langle \mathbf{y}, A(\mathbf{x}_k) + \mathbf{x}'_k \rangle - \langle \mathbf{c}, \mathbf{x}_k \rangle, \quad k \in \mathbb{N}.$$

Hence

$$\limsup_{k \to \infty} \langle \mathbf{c}, \mathbf{x}_k \rangle \leq \limsup_{k \to \infty} \langle \mathbf{y}, A(\mathbf{x}_k) + \mathbf{x}'_k \rangle = \lim_{k \to \infty} \langle \mathbf{y}, A(\mathbf{x}_k) + \mathbf{x}'_k \rangle = \langle \mathbf{y}, \mathbf{b} \rangle.$$

Since the feasible sequences were arbitrary, this means that the limit value of (P) is at most $\langle \mathbf{y}, \mathbf{b} \rangle$, and since $\mathbf{y}$ was an arbitrary feasible solution of (D), the lemma follows. $\qquad\square$

You may have expected the following stronger version of weak duality: *if both (P) and (D) are limit-feasible, then the limit value of (P) cannot exceed the limit value of (D)*. But a glance at Fig. 4.4 already shows that this is false in general.

Regular Duality

Here is the *regular Duality Theorem*. The proof essentially consists of applications of the Farkas lemma to carefully crafted systems.

4.7.3 Theorem. *The dual program (D) is feasible and has a finite value β if and only if the primal program (P) is limit-feasible and has a finite limit value γ. Moreover, $\beta = \gamma$.*

Proof. If (D) is feasible and has value β, we know that

$$A^T(\mathbf{y}) - \mathbf{c} \in K^*, \ \mathbf{y} \in L^* \quad \Rightarrow \quad \langle \mathbf{b}, \mathbf{y} \rangle \geq \beta. \qquad (4.13)$$

We also know that

$$A^T(\mathbf{y}) \in K^*, \ \mathbf{y} \in L^* \quad \Rightarrow \quad \langle \mathbf{b}, \mathbf{y} \rangle \geq 0. \qquad (4.14)$$

Indeed, if we had some $\mathbf{y}$ that fails to satisfy the latter implication, we could add a large positive multiple of it to any feasible solution of (D) and in this way obtain a feasible solution of value smaller than β.

We can now merge (4.13) and (4.14) into the single implication

$$A^T(\mathbf{y}) - z\mathbf{c} \in K^*, \ \mathbf{y} \in L^*, \ z \geq 0 \quad \Rightarrow \quad \langle \mathbf{b}, \mathbf{y} \rangle \geq z\beta. \qquad (4.15)$$

For $z > 0$, we obtain this from (4.13) by multiplication of all terms with z and then renaming $z\mathbf{y} \in L^*$ back to $\mathbf{y}$. For $z = 0$, it is simply (4.14). In matrix form as introduced on page 54, we can rewrite (4.15) as follows.

$$\begin{pmatrix} A^T & -\mathbf{c} \\ \hline \mathrm{id} & \mathbf{0} \\ \hline \mathbf{0} & 1 \end{pmatrix} (\mathbf{y}, z) \in K^* \oplus L^* \oplus \mathbb{R}_+ \quad \Rightarrow \quad \langle (\mathbf{b}, -\beta), (\mathbf{y}, z) \rangle \geq 0. \quad (4.16)$$

Here and in the following, we use a column vector $\mathbf{c} \in V$ as the linear operator $z \mapsto z\mathbf{c}$ from $\mathbb{R}$ to V and the row vector $\mathbf{c}^T$ as the (adjoint) linear operator $\mathbf{x} \mapsto \langle \mathbf{c}, \mathbf{x} \rangle$ from V to $\mathbb{R}$.

The matrix form (4.16) now allows us to apply the Farkas lemma: We are precisely in the situation where the second system has no solution. According to Lemma 4.5.6, the implication (4.16) holds if and only if the first system

$$\left(\begin{array}{c|c|c} A & \mathrm{id} & 0 \\ \hline -\mathbf{c}^T & \mathbf{0}^T & 1 \end{array} \right) (\mathbf{x}, \mathbf{x}', z) = (\mathbf{b}, -\beta), \ (\mathbf{x}, \mathbf{x}', z) \in (K^* \oplus L^* \oplus \mathbb{R}_+)^* = K \oplus L \oplus \mathbb{R}_+$$

$$(4.17)$$

is limit-feasible. The latter equality uses Lemma 4.3.3 together with Lemma 4.4.1.

System (4.17) is limit-feasible if and only if there are sequences $(\mathbf{x}_k)_{k \in \mathbb{N}}$, $(\mathbf{x}'_k)_{k \in \mathbb{N}}$, $(z_k)_{k \in \mathbb{N}}$ with $\mathbf{x}_k \in K$, $\mathbf{x}'_k \in L$, $z_k \geq 0$ for all k, such that

$$\lim_{k \to \infty} A(\mathbf{x}_k) + \mathbf{x}'_k = \mathbf{b} \quad (4.18)$$

and

$$\lim_{k \to \infty} \langle \mathbf{c}, \mathbf{x}_k \rangle - z_k = \beta. \quad (4.19)$$

Now (4.18) shows that (P) is limit-feasible, and (4.19) shows that the limit value of (P) is at least β. Weak duality (Theorem 4.7.2) shows that it is at most β, concluding the "only if" direction.

For the "if" direction, let (P) be limit-feasible with finite limit value γ and assume for the purpose of obtaining a contradiction that (D) is infeasible. This yields the implication

$$A^T(\mathbf{y}) - z\mathbf{c} \in K^*, \ \mathbf{y} \in L^*, \quad \Rightarrow \quad z \leq 0, \quad (4.20)$$

since for any pair $(\mathbf{y}, z)$ that violates it, $\frac{1}{z}\mathbf{y}$ would be a feasible solution of (D). We now play the same game as before and write this in Farkas-lemma-compatible matrix form (again, we have unsolvability of the second system):

$$\left(\begin{array}{c|c} A^T & -\mathbf{c} \\ \hline \mathrm{id} & 0 \end{array} \right) (\mathbf{y}, z) \in K^* \oplus L^* \quad \Rightarrow \quad \langle (\mathbf{0}, -1), (\mathbf{y}, z) \rangle \geq 0. \quad (4.21)$$

According to Lemma 4.5.6, this means that the first system

$$\left(\begin{array}{c|c} A & \mathrm{id} \\ \hline -\mathbf{c}^T & 0 \end{array}\right)(\mathbf{x}, \mathbf{x}') = (\mathbf{0}, -1), \quad (\mathbf{x}, \mathbf{x}') \in (K^* \oplus L^*)^* = K \oplus L \qquad (4.22)$$

is limit-feasible, which in turn means that there are sequences $(\mathbf{x}_k)_{k\in\mathbb{N}}$, $(\mathbf{x}'_k)_{k\in\mathbb{N}}$ with $\mathbf{x}_k \in K$, $\mathbf{x}'_k \in L$ for all k, such that

$$\lim_{k\to\infty} A(\mathbf{x}_k) + \mathbf{x}'_k = \mathbf{0} \qquad (4.23)$$

and

$$\lim_{k\to\infty} \langle \mathbf{c}, \mathbf{x}_k \rangle = 1. \qquad (4.24)$$

But this is a contradiction: elementwise addition of $(\mathbf{x}_k)_{k\in\mathbb{N}}$, $(\mathbf{x}'_k)_{k\in\mathbb{N}}$ to any feasible sequences of (P) that attain limit value γ would result in feasible sequences that witness limit value at least $\gamma + 1$.

Consequently, the dual program (D) must have been feasible. Weak duality (Theorem 4.7.2) yields that (D) has finite value $\beta \geq \gamma$. But then $\beta = \gamma$ follows from the previous "only if" direction. $\qquad\square$

Strong Duality

Here is the proof of the strong Duality Theorem of Cone Programming, under Slater's constraint qualification.

Proof of Theorem 4.7.1. The primal cone program (P) is feasible and hence, in particular, limit-feasible, and since it has an interior point, Theorem 4.6.5 shows that the limit value of (P) is equal to the value of (P) which is γ. Using the regular Duality Theorem 4.7.3 ("if" direction), the statement follows. $\qquad\square$

In Definition 4.6.4, the reader already may have wondered why a feasible point $\mathbf{x} \in \mathrm{int}(K)$ is called an interior point only in the case $L = \{\mathbf{0}\}$. The reason is that this case is indeed special. For general L, the existence of a feasible point in $\mathrm{int}(K)$ is not enough to guarantee that the value of (P) equals the limit value. Consequently, strong duality does not necessarily hold in such a situation.

To demonstrate this, we recall the example program on page 59 that has $K = \mathbb{R}^2$. It is feasible, has finite value 0 and a feasible solution in $\mathrm{int}(K) = K$. But the limit value is ∞. This also implies (by weak duality) that the dual program must be infeasible, and strong duality therefore fails in this example.

To gain a better understanding of the issue, the reader is invited to find the point in the proof of Theorem 4.6.5 where the condition $L = \{\mathbf{0}\}$ is crucial.

Semidefinite Programming Case

Using Theorem 4.7.1, we can now prove what we originally set out to prove, namely Theorem 4.1.1, the strong Duality Theorem of Semidefinite Programming. To this end, we apply Theorem 4.7.1 with $V = \mathrm{SYM}_n$, $W = \mathbb{R}^m$, $K = \mathrm{PSD}_n$ and $L = \{\mathbf{0}\}$. According to Lemma 4.5.3, the adjoint operator assumes the required form

$$A^T(\mathbf{y}) = \sum_{i=1}^{m} y_i A_i.$$

Generating matrices for the positive semidefinite cone. The last ingredient for the proof of Theorem 4.1.1 is to determine $(\mathrm{PSD}_n)^*$, the dual of the cone of all positive semidefinite matrices. As we will see, this cone happens to be self-dual, i.e., $(\mathrm{PSD}_n)^* = \mathrm{PSD}_n$.

First, we will show that PSD_n is generated by rank-one matrices; these will be easier to work with.

4.7.4 Lemma. *Let M be an $n \times n$ real matrix. We have $M \succeq 0$ if and only if there are unit-length vectors $\mathtt{B}_1, \ldots, \mathtt{B}_n \in S^{n-1}$ and nonnegative real numbers $\lambda_1, \ldots, \lambda_n$ such that*

$$M = \sum_{i=1}^{n} \lambda_i \mathtt{B}_i \mathtt{B}_i^T.$$

This result is, in a sense, analogous to the fact that the nonnegative orthant $\mathbb{R}_+^n$ can be characterized as the set of all linear combinations with nonnegative coefficients of the vectors $\mathbf{e}_1, \ldots, \mathbf{e}_n$ of the standard basis. However, in the case of the nonnegative orthant we have finitely many generators, but for PSD_n we need infinitely many generators.

Proof of Lemma 4.7.4. The "if" direction is immediate: the matrices $\mathtt{B}_i \mathtt{B}_i^T$ are positive semidefinite, and since PSD_n is a convex cone, every nonnegative linear combination of such matrices is in PSD_n as well.

For the "only if" direction, we diagonalize M as $M = SDS^T$, where S is orthogonal and D is the diagonal matrix of eigenvalues $\lambda_1, \ldots, \lambda_n$. The λ_i are nonnegative since $M \succeq 0$. Let $D^{(i)}$ be the diagonal matrix with λ_i at position (i, i) and zeros everywhere else. Then

$$M = S\left(\sum_{i=1}^{n} D^{(i)}\right)S^T = \sum_{i=1}^{n} SD^{(i)}S^T = \sum_{i=1}^{n} \lambda_i \mathtt{B}_i \mathtt{B}_i^T,$$

where $\mathtt{B}_i$ is the i-th column of S. By the orthogonality of S, $\|\mathtt{B}_i\| = 1$ for all i. $\square$

Now we are ready to prove the self-duality of PSD_n.

4.7.5 Lemma. $(\mathrm{PSD}_n)^* = \mathrm{PSD}_n$.

Proof. First we check that every $X \succeq 0$ also belongs to $(\mathrm{PSD}_n)^*$, which amounts to proving that $X \bullet Y \geq 0$ for every $X, Y \succeq 0$.

To this end, we write X in the form guaranteed by Lemma 4.7.4: $X = \sum_{i=1}^n \lambda_i \beta_i \beta_i^T$, where all $\lambda_i \geq 0$ and the β_i are unit vectors. Then we employ the expression of $X \bullet Y$ as $\mathrm{Tr}(X^T Y) = \mathrm{Tr}(XY)$, the linearity of the trace, and a "commutativity" property of it: $\mathrm{Tr}(AB) = \mathrm{Tr}(BA)$ (Exercise 4.4). We compute

$$X \bullet Y = \mathrm{Tr}\left(\sum_{i=1}^n \lambda_i \beta_i \beta_i^T Y \right) = \sum_{i=1}^n \lambda_i \mathrm{Tr}(\beta_i \beta_i^T Y)$$
$$= \sum_{i=1}^n \lambda_i \mathrm{Tr}(\beta_i^T Y \beta_i) = \sum_{i=1}^n \lambda_i \beta_i^T Y \beta_i \geq 0,$$

since $\beta_i^T Y \beta_i \geq 0$ by $Y \succeq 0$; see Fact 2.2.1(ii).

It remains to show the inclusion $(\mathrm{PSD}_n)^* \subseteq \mathrm{PSD}_n$. For this, we take an arbitrary $M \in (\mathrm{PSD}_n)^*$. For all $\mathbf{x} \in \mathbb{R}^n$, the matrix $\mathbf{x}\mathbf{x}^T$ is positive semidefinite, so using the same trick with the trace as above, from $M \in (\mathrm{PSD}_n)^*$ we derive

$$0 \leq M \bullet \mathbf{x}\mathbf{x}^T = \mathrm{Tr}((M\mathbf{x})\mathbf{x}^T) = \mathrm{Tr}(\mathbf{x}^T M \mathbf{x}) = \mathbf{x}^T M \mathbf{x}.$$

So we have $\mathbf{x}^T M \mathbf{x} \geq 0$ for all $\mathbf{x}$, and thus $M \succeq 0$. Lemma 4.7.5 is proved, and so is Theorem 4.1.1. $\square$

Figure 4.5 summarizes the whole route to the SDP duality theorem undertaken in this chapter.

4.8 The Largest Eigenvalue

Here we will re-prove a well-known fact from linear algebra using the duality of semidefinite programming. There exists a much shorter proof of this theorem, but we want to illustrate the use of duality in a very simple case. We will also prepare grounds for a more general statement; see Exercise 4.11.

4.8.1 Theorem. Let $C \in \mathrm{SYM}_n$. Then the largest eigenvalue of C is equal to

$$\lambda = \max\{\mathbf{x}^T C \mathbf{x} : \mathbf{x} \in \mathbb{R}^n, \|\mathbf{x}\| = 1\}.$$

We note that the maximum exists, since we are optimizing a continuous function $\mathbf{x}^T C \mathbf{x}$ over a compact subset of $\mathbb{R}^n$.

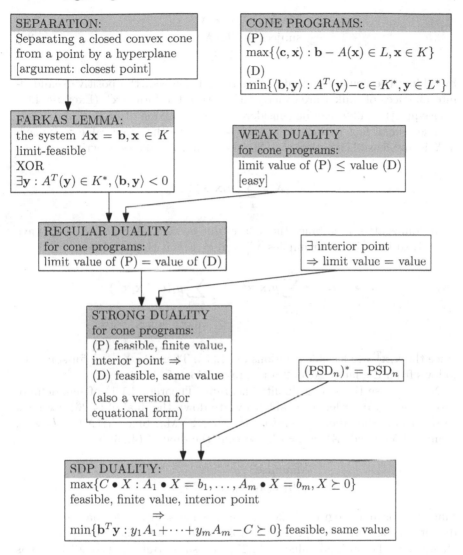

Fig. 4.5 The proof of the strong duality theorem for semidefinite programming

Proof. In the same spirit as in the proof of Lemma 4.7.5, we first rewrite $\mathbf{x}^T C \mathbf{x}$ as $C \bullet \mathbf{x}\mathbf{x}^T$ and $\|\mathbf{x}\| = 1$ as $\mathrm{Tr}(\mathbf{x}\mathbf{x}^T) = 1$. This means that λ is the value of the constrained optimization problem

$$
\begin{aligned}
\text{Maximize} \quad & C \bullet \mathbf{x}\mathbf{x}^T \\
\text{subject to} \quad & \mathrm{Tr}(\mathbf{x}\mathbf{x}^T) = 1.
\end{aligned}
\tag{4.25}
$$

This program is obtained from the semidefinite program

$$\text{Maximize} \quad C \bullet X$$
$$\text{subject to} \quad \text{Tr}(X) = 1 \qquad (4.26)$$
$$X \succeq 0$$

by adding the constraint that X has rank 1. Indeed, the positive semidefinite matrices of rank 1 are exactly the ones of the form $\mathbf{x}\mathbf{x}^T$ (Exercise 4.9). Consequently, (4.26) can be considered as a relaxation of (4.25).

The crucial fact is that (4.25) and (4.26) have the same value γ. Indeed, if X is any feasible solution of (4.26), Lemma 4.7.4 allows us write

$$X = \sum_{i=1}^n \mu_i \mathbf{x}_i \mathbf{x}_i^T,$$

with nonnegative μ_i's. Since the $\mathbf{x}_i$ are unit vectors, all matrices $\mathbf{x}_i \mathbf{x}_i^T$ have trace 1, so linearity of Tr implies $\sum_{i=1}^n \mu_i = 1$. But then we have

$$C \bullet X = C \bullet \sum_{i=1}^n \mu_i \mathbf{x}_i \mathbf{x}_i^T \;=\; \sum_{i=1}^n \mu_i (C \bullet \mathbf{x}_i \mathbf{x}_i^T)$$
$$\leq \; \max_{i=1}^n C \bullet \mathbf{x}_i \mathbf{x}_i^T \leq \gamma,$$

since the $\mathbf{x}_i \mathbf{x}_i^T$ are feasible solutions of (4.25). Then $C \bullet X^* \geq \lambda$ for some X^* follows from the fact that (4.26) is a relaxation of (4.25).

Next we use the strong duality theorem (Theorem 4.1.1) of semidefinite programming. In order to be able to write down the dual of (4.26), we need to determine the adjoint operator $A^T \colon \mathbb{R} \to \text{SYM}_n$. Since $\text{Tr}(X) = I_n \bullet X$, Lemma 4.5.3 yields $A^T(y) := yI_n$, so that the dual of (4.26) is

$$\text{Minimize} \quad y$$
$$\text{subject to} \quad yI_n - C \succeq 0. \qquad (4.27)$$

Since the primal program (4.26) has a feasible solution X that is positive definite (choose $X = \frac{1}{n}I_n$, for example), the duality theorem applies and shows that the optimal value of (4.27) is also λ. But what *is* λ? If C has eigenvalues $\lambda_1, \ldots, \lambda_n$, then $yI_n - C$ has eigenvalues $y - \lambda_1, \ldots, y - \lambda_n$, and the constraint $yI_n - C \succeq 0$ requires all of them to be nonnegative. Therefore, the value λ, the smallest y for which $yI_n - C \succeq 0$ holds, equals the largest eigenvalue of C. This proves the theorem. $\qquad\qquad\square$

Exercise 4.11 discusses semidefinite programming formulations for the sum of the k largest eigenvalues.

Exercises

4.1 Let
$$\triangleleft' = \{(x, y, z) \in \mathbb{R}^3 : x \geq 0, y \geq 0, 2xy \geq z^2\}$$
be a vertically stretched version of the toppled ice cream cone $\triangleleft$. Prove that there is an orthogonal matrix M (i.e., $M^{-1} = M^T$) such that

$$\triangleleft' = \{M\mathbf{x} : \mathbf{x} \in \nabla_3\}.$$

This means that the toppled ice cream cone is an isometric image of the ice cream cone, plus some additional vertical squeezing (which naturally happens when soft objects topple over).

To understand why the isometric image indeed corresponds to "toppling," analyze what M does to the "axis" $\{(0, 0, r) \in \mathbb{R}^3 : r \in \mathbb{R}_+\}$ of the ice cream cone.

4.2 What is the dual of the ice cream cone ∇_n (see page 48)?

4.3 Prove Lemma 4.3.3: The dual of a direct sum of cones K and L is the direct sum of the dual cones K^* and L^*.

4.4 Let $A \in \mathbb{R}^{m \times n}$, $B \in \mathbb{R}^{n \times m}$ be two rectangular matrices. (Then AB and BA are both square matrices.) Prove that

$$\mathrm{Tr}(AB) = \mathrm{Tr}(BA).$$

4.5 Prove or disprove the following claim:

$$\mathrm{Tr}(ABC) = \mathrm{Tr}(ACB)$$

for all square matrices A, B, C. If you conclude that the claim is false in general, what about the case where $A, B, C \in \mathrm{SYM}_n$?

4.6 Let γ' be the limit value of a limit-feasible cone program

$$\begin{aligned} \text{Maximize} \quad & \langle \mathbf{c}, \mathbf{x} \rangle \\ \text{subject to} \quad & \mathbf{b} - A(\mathbf{x}) \in L \\ & \mathbf{x} \in K. \end{aligned}$$

Prove that there is a feasible sequence $(\mathbf{x}_k)_{k \in \mathbb{N}}$ such that

$$\gamma' = \limsup_{k \to \infty} \langle \mathbf{c}, \mathbf{x}_k \rangle.$$

4.7 Prove the following statement that introduces yet another kind of constraint qualification.

If the primal program (P) is feasible, has finite value γ *and* the cone

$$C = \left\{ \left(\begin{array}{c|c} A & \mathrm{id} \\ \hline \mathbf{c}^T & \mathbf{0}^T \end{array} \right) (\mathbf{x}, \mathbf{x}') : (\mathbf{x}, \mathbf{x}') \in K \times L \right\}$$

is closed, then the dual program (D) is feasible and has finite value $\beta = \gamma$. (We again refer to the matrix notation of operators introduced on page 54.)

We remark that if K and L are "linear programming cones" (meaning that they are iterated direct sums of one-dimensional cones $\{0\}, \mathbb{R}, \mathbb{R}^+$), then the above cone C is indeed closed, see e.g., [GM07, Chap. 6.5]. It follows that strong linear programming duality requires no constraint qualification.

Hint: First reprove Theorem 4.6.5 under the given constraint qualification.

4.8 Let $M \in \mathrm{SYM}_n$ be a symmetric $n \times n$ matrix with eigenvalues $\lambda_1, \ldots, \lambda_n$. Prove that $\mathrm{Tr}(M) = \sum_{j=1}^{n} \lambda_j$.

4.9 Prove that a matrix $M \in \mathrm{SYM}_n$ has rank 1 if and only if $M = \pm \beta\beta^T$ for some nonzero vector $\beta \in \mathbb{R}^n$. In particular, $M \succeq 0$ has rank 1 if and only if $M = \beta\beta^T$.

4.10 Given a symmetric matrix $C \in \mathrm{SYM}_n$, we are looking for a matrix $Y \succeq 0$ such that $Y - C \succeq 0$. Prove that the trace of every such matrix is at least the sum of the positive eigenvalues of C. Moreover, there exists a matrix $\tilde{Y} \succeq 0$ with $\tilde{Y} - C \succeq 0$ and $\mathrm{Tr}(\tilde{Y})$ equal to the sum of the positive eigenvalues of C.

4.11 Let $C \in \mathrm{SYM}_n$.

(a) Prove that the value of the following cone program is the sum of the k largest eigenvalues of C.

$$\begin{array}{rl} \text{Minimize} & ky + Tr(Y) \\ \text{subject to} & yI_n + Y - C \succeq 0 \\ & (Y, y) \in \mathrm{PSD}_n \oplus \mathbb{R}. \end{array}$$

Hint: You may use the statement of Exercise 4.10

(b) Derive the dual program and show that its value is also the sum of the k largest eigenvalues of C. In doing this, you have (almost) proved *Fan's Theorem*: The sum of the k largest eigenvalues of C is the equal to

$$\max C \bullet YY^T,$$

where Y ranges over all $n \times k$ matrices such that $Y^T Y = I_k$.

4.12 For a given graph $G = (V, E)$ with $V = \{1, 2, \ldots, n\}$, consider the semidefinite program

$$\begin{aligned}
\text{maximize} \quad & J_n \bullet X \\
\text{subject to} \quad & \text{Tr}(X) = 1 \\
& x_{ij} = 0, \ \{i,j\} \in E \\
& X \succeq 0,
\end{aligned}$$

where J_n is the all-one $n \times n$ matrix, and show that its value is $\vartheta(G)$ (see Chap. 3).

Hint: Dualize the semidefinite program in Theorem 3.6.1.

4.13 Prove that Lemma 3.4.2 actually holds with equality; that is,

$$\vartheta(G \cdot H) \geq \vartheta(G)\vartheta(H)$$

for all graphs G, H.

Hint: Use the expression of $\vartheta(G)$ from Exercise 4.12.

Chapter 5
Approximately Solving Semidefinite Programs

In this chapter we present an algorithm for solving a certain class of semidefinite programs in polynomial time, up to any desired accuracy.

As we mentioned in Sect. 2.6, no algorithm is known that solves *every* semidefinite program in time polynomial in the input size (even the representation of the output can be problematic – we recall that there are semidefinite programs for which the coordinates of all feasible points have exponentially many digits). Each of the known algorithms provides only an approximate solution, and requires some restriction on the input instances.

The theoretically strongest (but quite impractical) algorithm, the ellipsoid method, runs in time polynomial in the input size and in $\log(R/\varepsilon)$, where ε is the maximum allowed error and R is an upper bound on the norm of all feasible solutions for a given semidefinite program. Moreover, "polynomial time" refers to the bit model of computation (or equivalently, the Turing machine model).

The algorithm discussed in this chapter works only with a somewhat restricted class of semidefinite programs: namely, optimizing a linear function, subject to linear constraints, over all positive semidefinite matrices *of trace* 1, which is a substantial restriction. On the other hand, every semidefinite program with an a priori upper bound on the trace of some optimal solution can easily be reduced to this form; see Sect. 5.2. Such *trace-bounded* semidefinite programs cover many applications, including most of those in combinatorial optimization. For an illustrative example we come back to the MAXCUT problem on page 79 below.

The running time of the algorithm depends polynomially on $\frac{1}{\varepsilon}$, rather than on $\log\frac{1}{\varepsilon}$. Finally, in order to avoid a tedious analysis of rounding errors etc., we establish the polynomial bound on the running time in the *real RAM* model of computation, as opposed to the bit model (see page 5), i.e., we count arithmetic operations with real numbers.

On the other hand, the algorithm is simple (compared to other SDP algorithms), easy to implement, and quite efficient in practice. Moreover, its

B. Gärtner and J. Matoušek, *Approximation Algorithms and Semidefinite Programming*, DOI 10.1007/978-3-642-22015-9_5,
© Springer-Verlag Berlin Heidelberg 2012

analysis is also reasonably simple – we will be able to bound the running time without too much technical effort.

The algorithm originated in a 1956 paper by Frank and Wolfe [FW56], which dealt with convex quadratic programs. It was adapted for semidefinite programming by Hazan [Haz08], and we will refer to it as *Hazan's algorithm*. (More on its history will be mentioned in Sect. 5.4.)

Another, faster but more involved, recent algorithm for approximately solving semidefinite programs is due to Arora and Kale [AK07, Kal08]. It is based on the *multiplicative weights update* method that also appears in many other applications [AHK05].

5.1 Optimizing Over the Spectahedron

Hazan's algorithm deals with semidefinite programs of the form

$$
\begin{aligned}
\text{maximize} \quad & C \bullet X \\
\text{subject to} \quad & A_i \bullet X \leq b_i, \quad i = 1, 2, \ldots, m \\
& \text{Tr}(X) = 1 \\
& X \succeq 0.
\end{aligned}
\tag{5.1}
$$

Without loss of generality, we assume that all the A_i are symmetric and nonzero $n \times n$ matrices.

The essential difference to a general semidefinite program as introduced in Sect. 2.4 is the constraint $\text{Tr}(X) = 1$. With this constraint present, we may consider (5.1) as a "linear program" over the *spectahedron*[1]

$$
\text{Spect}_n := \{X \in \text{SYM}_n : \text{Tr}(X) = 1, X \succeq 0\}.
\tag{5.2}
$$

The spectahedron is a compact set, a "semidefinite analog" of the simplex. Hence the feasible region is bounded, and we always have an optimal solution if the problem is feasible at all.

The performance of Hazan's algorithm depends on m, n, the desired relative error ε, and the input size, which we measure by the quantity

$$
N := n + \text{supp}(C) + \sum_{i=1}^{m} \text{supp}(A_i),
$$

where $\text{supp}(A)$ is the number of nonzero entries of a matrix A.

Here is what we will prove (recall that the Frobenius norm $\|A\|_F$ of a matrix is defined as $\|A\|_F = \sqrt{A \bullet A} = \sqrt{\sum_{ij} a_{ij}^2}$).

---

[1] Not to be confused with a *spectrahedron*, which is an arbitrary intersection of the cone of positive semidefinite matrices with an affine subspace of SYM_n.

5.1.1 Theorem. *Let γ_{opt} be the value of the semidefinite program (5.1), and let $\varepsilon \in (0, 1]$ be given. Hazan's algorithm either finds a matrix $\tilde{X} \in \mathrm{Spect}_n$ such that*

$$\frac{\gamma_{\mathrm{opt}} - C \bullet \tilde{X}}{\|C\|_F} \leq 2\varepsilon, \qquad \frac{A_i \bullet \tilde{X} - b_i}{\|A_i\|_F} \leq \varepsilon, \quad i = 1, 2, \ldots, m,$$

or correctly reports that (5.1) has no feasible solution. The number of arithmetic operations is bounded by

$$O\left(\frac{N(\log n)(\log m)\log(1/\varepsilon)}{\varepsilon^3} \right)$$

with high probability, assuming that $1/\varepsilon$ is bounded by some fixed polynomial function of n.

Here and in the subsequent theorems, "high probability" means probability at least $1 - 1/n^c$ for an arbitrary but fixed $c > 0$. The probability is with respect to internal random choices made by the algorithm (so we do not assume any kind of random input; the result holds for *every* input).

Eigenvalue computations. As we will see, Hazan's algorithm needs a subroutine for approximate computation of the largest eigenvalue of a positive semidefinite matrix. It solves the semidefinite program (5.1) using

$$O\left(\frac{(\log m)\log(1/\varepsilon)}{\varepsilon^2} \right)$$

invocations of that subroutine (plus auxiliary computations, which altogether take only $O(\frac{N(\log m)\log(1/\varepsilon)}{\varepsilon^2})$ operations).

Computing the largest eigenvalue of a matrix is a well-studied problem in numerical analysis, and several efficient methods are known. We have chosen a very simple and well-known one, the *power method*, which is also reasonably efficient. It is in this method where we need randomization; otherwise, Hazan's algorithm is deterministic.

For completeness, we also present a worst-case analysis of the performance of the power method in our setting (Sect. 5.8). We will need to apply the power method only for positive definite matrices, and then, with high probability, it requires no more than $O(\frac{N\log n}{\varepsilon})$ arithmetic operations.

This is really a worst-case bound. For "typical" matrices, the power method is much faster. Namely, if the largest eigenvalue of the considered matrix is well separated from the second largest one (which is the case for "most" matrices), then $O(N\log\frac{n}{\varepsilon})$ arithmetic operations suffice. One can thus expect that for typical inputs, Hazan's algorithm is also much faster than the worst-case analysis indicates.

The power method could also be replaced by other approaches. Perhaps most notably, one can use the *Lanczos algorithm* [GvL96, Chap. 9]. In its exact version, the Lanczos algorithm performs n iterations to transform the given matrix into a tridiagonal matrix (a matrix $T = (t_{ij})$ is *tridiagonal* if $t_{ij} = 0$ whenever $|i - j| \geq 2$). Since there is a simple recursive formula for the characteristic polynomial of a tridiagonal matrix, the largest eigenvalue can be found quickly through simple interval bisection.

The approximate version of the Lanczos algorithm performs $k < n$ iterations and outputs the largest eigenvalue of the $k \times k$ tridiagonal matrix T_k obtained at this point. In our setting, this approximate computation of the largest eigenvalue requires $O(N \frac{\log n}{\sqrt{\varepsilon}})$ arithmetic operations with high probability [KW92].

5.2 The Case of Bounded Trace

Before explaining Hazan's algorithm, we will show how the rather stringent requirement $\mathrm{Tr}(X) = 1$ can be relaxed. In a nutshell, we will see that it can be replaced by $\mathrm{Tr}(X) \leq t$ for a parameter t, but the price to pay is a polynomial dependence of the running time on t. This part is quite simple, but since it may be important for applications, we discuss it explicitly nevertheless.

A trace-bounded semidefinite program is of the form

$$
\begin{aligned}
\text{maximize} \quad & C \bullet X \\
\text{subject to} \quad & A_i \bullet X \leq b_i, \quad i = 1, 2, \ldots, m, \\
& \mathrm{Tr}(X) \; \{\leq, =\} \; t \\
& X \succeq 0,
\end{aligned}
\tag{5.3}
$$

where t is some positive real number, and $\{\leq, =\}$ stands for either an inequality or an equality constraint. The algorithm in Theorem 5.1.1 can also be used to solve trace-bounded semidefinite programs.

First, if the constraint on the trace is the inequality $\mathrm{Tr}(X) \leq t$, we can add an extra row and column of slack variables to X, obtaining a new matrix X'. Then the constraint $\mathrm{Tr}(X) \leq t$ is replaced with $\mathrm{Tr}(X') = t$.

So, assuming that we have the constraint $\mathrm{Tr}(X) = t$ in (5.3), we can substitute $X = tY$, $b_i = b_i't$, and we obtain an equivalent semidefinite program for which Theorem 5.1.1 applies (with ε/t in the role of ε). Thus, we obtain the following.

5.2.1 Corollary. *Let γ_{opt} be the value of the semidefinite program (5.3), and let $\varepsilon \in (0, 1]$ be given. Hazan's algorithm either finds a matrix $\tilde{X} \succeq 0$ such that $\mathrm{Tr}(\tilde{X}) \; \{\leq, =\} \; t$ and*

$$
\frac{\gamma_{\mathrm{opt}} - C \bullet \tilde{X}}{\|C\|_F} \leq 2\varepsilon, \quad \frac{A_i \bullet \tilde{X} - b_i}{\|A_i\|_F} \leq \varepsilon, \quad i = 1, 2, \ldots, m,
$$

or correctly reports that (5.3) has no feasible solution. The number of arithmetic operations is bounded by

$$O\left(\frac{N(\log n)(\log m)t^3 \log(t/\varepsilon)}{\varepsilon^3}\right)$$

with high probability, assuming that t/ε is bounded by some fixed polynomial function of n.

Solving the Semidefinite Relaxation of MaxCut

As an illustration, let us see what Corollary 5.2.1 gives us for the semidefinite relaxation (1.3) of the MaxCut problem:

$$\begin{array}{ll} \text{maximize} & \sum_{\{i,j\}\in E} \frac{1-x_{ij}}{2} \\ \text{subject to} & x_{ii} = 1, \quad i = 1, 2, \ldots, n \\ & X \succeq 0. \end{array}$$

Here $G = (\{1, 2, \ldots, n\}, E)$ is the graph whose maximum cut we want to approximate.

5.2.2 Theorem. *Let $\mathsf{SDP}(G)$ be the maximum value of the semidefinite relaxation of the MaxCut problem, for a given graph $G = (V, E)$ with n vertices. Given a constant $\varepsilon \in (0, \frac{1}{2})$, we can compute a matrix $\tilde{X} \succeq 0$ such that*

(i) $|\tilde{x}_{ii} - 1| \leq \varepsilon$ for $i = 1, 2, \ldots, n$

(ii) $\sum_{\{i,j\}\in E} \frac{1-\tilde{x}_{ij}}{2} \geq \mathsf{SDP}(G) - \varepsilon\sqrt{|E|}$.

With high probability, this requires

$$O\left((n^4 + n^3|E|)\log^3 n\right)$$

arithmetic operations, with a hidden cubic dependence on $1/\varepsilon$.

Since $\mathsf{SDP}(G) \geq |E|/2$, the relative error made in approximately solving the semidefinite relaxation of the MaxCut problem is at most $2\varepsilon/\sqrt{|E|}$. Actually, we can even require that $\tilde{x}_{ii} = 1$ for all i, which means that we get a *feasible* approximately optimal solution (see Exercise 5.1).

Proof. The objective function of the MaxCut relaxation can be written as

$$\frac{|E|}{2} - \frac{A_G \bullet X}{4},$$

where A_G is the adjacency matrix of G (the $n \times n$ matrix with $a_{ij} = 1$ if $\{i, j\}$ is an edge and $a_{ij} = 0$ otherwise). After removing the constant term, and replacing each equality constraint with a pair of inequality constraints, we get a trace-bounded problem in the form (5.3), with maximum value $\gamma_{\text{opt}} = \text{SDP}(G) - |E|/2$:

$$\begin{array}{ll}
\text{maximize} & -A_G \bullet X/4 \\
\text{subject to} & x_{ii} \leq 1, \quad i = 1, 2, \ldots, n \\
& x_{ii} \geq 1, \quad i = 1, 2, \ldots, n \\
& \text{Tr}(X) = n \\
& X \succeq 0.
\end{array}$$

We have $m = 2n$, $N = O(n + |E|)$, and $t = n$. The algorithm of Corollary 5.2.1 then has the claimed runtime and produces a matrix $\tilde{X} \succeq 0$ such that $\text{Tr}(\tilde{X}) = n$ and

$$\frac{\gamma_{\text{opt}} + A_G \bullet \tilde{X}/4}{\|A_G\|_F/4} \leq 2\varepsilon, \quad |\tilde{x}_{ii} - 1| \leq \varepsilon, \ i = 1, 2, \ldots, n.$$

Since $\|A_G\|_F = \sqrt{2|E|} < 2\sqrt{|E|}$, this implies

$$\sum_{\{i,j\}\in E} \frac{1 - \tilde{x}_{ij}}{2} = \frac{|E|}{2} - \frac{A_G \bullet \tilde{X}}{4} \geq \frac{|E|}{2} + \gamma_{\text{opt}} - \varepsilon\sqrt{|E|} = \text{SDP}(G) - \varepsilon\sqrt{|E|},$$

as claimed. $\square$

The runtime bound of Theorem 5.2.2 is polynomial but by far not the best known. Klein and Lu showed in 1996 [KL96] that the semidefinite relaxation of the MAXCUT problem can approximately be solved in time $\tilde{O}(n|E|)$, where the $\tilde{O}$ notation suppresses polylogarithmic factors. This is still the best known general bound. For graphs whose maximum degree is larger than the average degree only by a constant factor, the multiplicative weights update method of Arora and Kale yields an algorithm with runtime $\tilde{O}(|E|)$ [Kal08, Theorem 14].

5.3 The Semidefinite Feasibility Problem

Here we start with the presentation of Hazan's algorithm. We first reduce the semidefinite program (5.1) in a standard way to a sequence of *feasibility problems* of the following form:

$$\begin{array}{lll}
\text{find} & X \in \text{SYM}_n & \\
\text{subject to} & A_i \bullet X \leq b_i, & i = 1, 2, \ldots, m, \\
& \text{Tr}(X) = 1 & \\
& X \succeq 0.
\end{array} \tag{5.4}$$

We also assume (by rescaling the inequalities) that $\|A_i\|_F = 1$ for all $i = 1, 2, \ldots, m$, which will simplify the analysis (it allows us to switch to an additive error bound).

5.3.1 Definition. *Let $\varepsilon > 0$ be a real number. A matrix $X \in \mathrm{Spect}_n$ such that*

$$A_i \bullet X - b_i \leq \varepsilon, \quad i = 1, 2, \ldots, m$$

is called an ε-approximate solution of the feasibility problem (5.4).

Here is our main result, whose proof will appear in Sect. 5.6 below. This time we measure the input size by the parameter

$$N := n + \sum_{i=1}^{m} \mathrm{supp}(A_i).$$

5.3.2 Theorem. *There is an algorithm that, given an instance of the feasibility problem (5.4) and a number $\varepsilon \in (0, 1]$, either finds an ε-approximate solution, or correctly reports that the feasibility problem has no solution. The number of arithmetic operations required by the algorithm is bounded by*

$$O\left(\frac{N(\log n)(\log m)}{\varepsilon^3}\right),$$

with high probability.

To derive Theorem 5.1.1 from Theorem 5.3.2, we want to find an approximate solution of a semidefinite program of the form (5.1) by approximately solving a sequence of feasibility problems. This is based on the following fact, whose straightforward proof we omit.

5.3.3 Fact. *Let $\gamma \in \mathbb{R}$. The semidefinite program (5.1) has a solution of value at least γ if and only if the feasibility problem*

$$
\begin{aligned}
&\text{Find} && X \in \mathrm{SYM}_n \\
&\text{subject to} && C' \bullet X \geq \gamma' \\
& && A_i' \bullet X \leq b_i', \quad i = 1, 2, \ldots, m, \\
& && \mathrm{Tr}(X) = 1 \\
& && X \succeq 0
\end{aligned}
\tag{5.5}
$$

has a feasible solution, where

$$C' := \frac{C}{\|C\|_F}, \quad \gamma' := \frac{\gamma}{\|C\|_F}, \quad A_i' := \frac{A_i}{\|A_i\|_F}, \quad b_i' := \frac{b_i}{\|A_i\|_F}.$$

The strategy is now to perform a binary search for the largest value of γ' for which the algorithm in Theorem 5.3.2 still finds an ε-approximate solution $\tilde{X}$ of (5.5). We start with the following lemma.

5.3.4 Lemma. *Let $C \in \mathrm{SYM}_n$, and $X \in \mathrm{Spect}_n$. Then $|C \bullet X| \leq \|C\|_F$.*

Proof. By the Cauchy–Schwarz inequality, we have $|C \bullet X| \leq \|C\|_F \|X\|_F$, and it is easy to show that $\|X\|_F \leq 1$ for every $X \in \mathrm{Spect}_n$ (Exercise 5.5). $\qquad\square$

The lemma implies that the search space for γ' can be restricted to the interval $[-1, 1]$ where we can apply binary search. The following easy result, whose proof we again omit, then yields Theorem 5.1.1.

5.3.5 Lemma. *Suppose that the semidefinite program (5.1) is feasible. Let γ_{opt} be its optimum value, and let $\varepsilon \in (0, 1]$. By solving at most $\log \frac{1}{\varepsilon}$ feasibility problems of the form (5.5), we can find a value $\gamma' \geq \gamma_{\mathrm{opt}}/\|C\|_F - \varepsilon$ and an ε-approximate solution $\tilde{X}$ of (5.5) for that value of γ'. The matrix $\tilde{X}$ satisfies the requirements of Theorem 5.1.1.* $\qquad\square$

In order to maintain "high probability" throughout this reduction, we need the (mild) assumption that $\log(1/\varepsilon)$ is bounded by a polynomial in n.

5.4 Convex Optimization Over the Spectahedron

We now turn to the problem of finding an ε-approximate solution of the feasibility problem (5.4). Instead of trying to satisfy the linear constraints $A_i \bullet X \leq b_i$ directly, we are going to solve a convex minimization problem over the spectahedron Spect_n, where the convex objective function f_{pen} is chosen so that it *penalizes* violations of the constraints.[2] Namely, we solve the problem

$$\text{minimize } f_{\mathrm{pen}}(X) \text{ subject to } X \in \mathrm{Spect}_n, \qquad (5.6)$$

where

$$f_{\mathrm{pen}}(X) := \frac{1}{K} \log \left(\sum_{i=1}^{m} e^{K(A_i \bullet X - b_i)} \right), \quad K := \frac{2 \log m}{\varepsilon}. \qquad (5.7)$$

Checking the convexity of f_{pen} is left as Exercise 5.2.

Hazan's algorithm will find an approximate solution $\tilde{X}$ of the convex optimization problem (5.6) that is optimal up to an additive error of at most $\varepsilon/2$. The next lemma shows that such an $\tilde{X}$ either is the desired ε-approximate solution of the feasibility problem (5.4), or it certifies infeasibility.

---

[2] A function f is convex over a set D if for all $\mathbf{x}, \mathbf{y} \in D$ and $\lambda \in [0, 1]$, we have $f((1 - \lambda)\mathbf{x} + \lambda\mathbf{y}) \leq (1 - \lambda)f(\mathbf{x}) + \lambda f(\mathbf{y})$.

5.4.1 Lemma. Let X^* be an optimal solution of (5.6), and let $\tilde{X} \in \mathrm{Spect}_n$ be such that $f_{\mathrm{pen}}(\tilde{X}) \le f_{\mathrm{pen}}(X^*) + \varepsilon/2$. Then the following statements hold.

(i) If $f_{\mathrm{pen}}(\tilde{X}) \le \varepsilon$, then $\tilde{X}$ is an ε-approximate solution of the feasibility problem (5.4).

(ii) If $f_{\mathrm{pen}}(\tilde{X}) > \varepsilon$, then the feasibility problem has no solution.

Proof. For our choice of $K = 2\log m/\varepsilon$, it is easy to calculate (see Exercise 5.3) that

$$\max_i(A_i \bullet X - b_i) \le f_{\mathrm{pen}}(X) \le \max_i(A_i \bullet X - b_i) + \varepsilon/2 \text{ for all } X \in \mathbb{R}^{n \times n}.$$

Hence if $f_{\mathrm{pen}}(\tilde{X}) \le \varepsilon$, then the lower bound shows that $\tilde{X}$ is an ε-approximate solution.

On the other hand, for $f_{\mathrm{pen}}(\tilde{X}) > \varepsilon$ we know that for all $X \in \mathrm{Spect}_n$, we have $\varepsilon/2 < f_{\mathrm{pen}}(X) \le \max_i(A_i \bullet X - b_i) + \varepsilon/2$, meaning that there is no feasible solution. $\qquad\square$

The desired approximate minimum of f_{pen} is obtained by the *Frank–Wolfe algorithm*. Originally, the algorithm was developed in 1956 to solve convex quadratic programs through a sequence of linear programs [FW56]. The algorithm works for general convex programs, and this was recently used by Clarkson [Cla10], who designed efficient approximation algorithms for a number of mostly geometric optimization problems, based on the Frank–Wolfe algorithm. Clarkson's approach deals with convex optimization problems whose feasible region is a simplex. Hazan showed that it can also be used if the feasible region is the spectahedron.

Gradients. The Frank–Wolfe algorithm works with the gradient of the minimized function, so we set up some conventions and state a simple fact concerning this gradient.

For a function $f: \mathbb{R}^{n \times n} \to \mathbb{R}$, it is customary to write gradients as matrices. Formally, we define $\nabla f(X)$ as the matrix $D = (d_{ij})$ with

$$d_{ij} = \frac{\partial f}{\partial x_{ij}}(X), \quad 1 \le i, j \le n.$$

If f is symmetric, meaning that $f(X) = f(X^T)$, it follows that $\nabla f(X) \in \mathrm{SYM}_n$ for all $X \in \mathrm{SYM}_n$. Moreover, this definition of gradient is compatible with scalar products in the sense that

$$\nabla f(X) \bullet Y = \langle \nabla \mathbf{f}(\mathbf{X}), \mathbf{Y} \rangle,$$

where $\mathbf{f}$, $\mathbf{X}$, $\mathbf{Y}$ are obtained from f, X, and Y by identifying $R^{n \times n}$ with R^{n^2} in a canonical way (listing the matrix entries row by row).

The proof of the following lemma follows from standard calculus and we omit it.

5.4.2 Lemma. *The penalty function* f_{pen} *as defined in (5.7) is symmetric, and*

$$\nabla f_{\text{pen}}(X) = \frac{\sum_{i=1}^{m} z_i(X) A_i}{\sum_{i=1}^{m} z_i(X)} \in \text{SYM}_n,$$

where

$$z_i(X) := e^{K(A_i \bullet X - b_i)}, \quad i = 1, 2, \ldots, m. \qquad \square$$

This means that $\nabla f_{\text{pen}}(X)$ is a convex combination of the matrices A_i, where the coefficient of A_i is exponential in the "amount of violation" $A_i \bullet X - b_i$.

5.5 The Frank–Wolfe Algorithm

In this section we describe the algorithm for minimizing f_{pen} over the spectahedron. The algorithm can actually handle any function f that satisfies the following condition.

5.5.1 Assumption. $f \colon \mathbb{R}^{n \times n} \to \mathbb{R}$ *is a convex and symmetric function. Furthermore, all first and second partial derivatives of* f *are continuous.*

So we consider the optimization problem

$$\text{minimize } f(X) \text{ subject to } X \in \text{Spect}_n, \tag{5.8}$$

with f as in Assumption 5.5.1. Our goal is to find a matrix $\tilde{X} \in \text{Spect}_n$ such that

$$f(\tilde{X}) \leq f(X^*) + \varepsilon/2,$$

where X^* is an exact minimizer.

The algorithm is of gradient-based descent type. We start from some initial matrix $X_1 \in \text{Spect}_n$, and we construct a sequence $X_2, X_3, \ldots$ of matrices in Spect_n.

Given X_k, we first solve a linearized version of (5.8) to get a descent direction (it need not be the *steepest* descent direction, as in some other gradient-based minimization algorithms), and then we make a step in this direction to obtain X_{k+1}. One of the main twists in the algorithm is a careful choice of the length of the step.

Using elementary multivariate calculus, we get that the equation of the tangent hyperplane to the graph of f at the point $(X_k, f(X_k))$ is

$$\ell(X) = f(X_k) + \nabla f(X_k) \bullet (X - X_k).$$

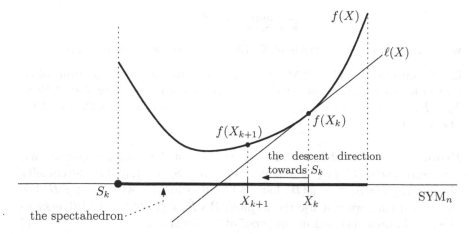

Fig. 5.1 A schematic illustration of a typical step of the Frank–Wolfe algorithm

Thus, the linearized problem (the Frank–Wolfe linearization) is to minimize $\ell(X)$ over the spectahedron. (For the purpose of this minimization, we can remove the constant term of $\ell(X)$, which we do in the description of the algorithm below.)

It is good to realize that the minimum of $\ell(X)$ is always attained on the boundary of Spect_n, and its location may thus be nowhere close to the desired minimizer for $f(X)$. But in the algorithm, we use the minimizer of $\ell(X)$ only for finding a local "descent direction" at X_k, and we do not expect it to approximate the minimizer of $f(X)$.

The Frank–Wolfe algorithm proceeds as follows.

5.5.2 Algorithm (Frank–Wolfe on the spectahedron).

1. Set $X_1 := \mathbf{x}\mathbf{x}^T$ for an arbitrary unit vector $\mathbf{x} \in \mathbb{R}^n$.
2. For $k = 1, 2, \ldots$ do:

   2.1 Compute a matrix $S_k \in \mathrm{Spect}_n$ such that

   $$\nabla f(X_k) \bullet S_k \leq \min_{X \in \mathrm{Spect}_n} \nabla f(X_k) \bullet X + \frac{\varepsilon}{4}.$$

   2.2 Set $X_{k+1} := X_k + \alpha_k (S_k - X_k)$, where $\alpha_k := \min\left(1, \frac{2}{k}\right)$.

Step 2 is illustrated in Fig. 5.1.

This formulation of the algorithm does not yet tell us how many iterations we need to perform; we will derive a bound on this below; see page 88.

Finally, we need to specify how Step 2.1 computes an approximate solution of the linearized problem

$$\min_{X \in \mathrm{Spect}_n} B \bullet X,$$

where $B \in \mathrm{SYM}_n$ is the gradient $\nabla f(X_k)$. Here is the crucial lemma.

5.5.3 Lemma. *Let $B \in \mathrm{SYM}_n$, let $\lambda_{\min}(B)$ be the smallest eigenvalue of B, and let β be a δ-approximate smallest eigenvector of B, meaning that $\beta^T B \beta \le \lambda_{\min}(B) + \delta$. Then $\min_{X \in \mathrm{Spect}_n} B \bullet X = \lambda_{\min}(B)$, and for $\tilde{X} := \beta \beta^T$, we have $B \bullet \tilde{X} \le \lambda_{\min}(B) + \delta$.*

Proof. In the proof of Theorem 4.8.1 we saw that the largest eigenvalue of a symmetric matrix C is equal to $\max\{C \bullet X : X \in \mathrm{Spect}_n\}$, and so the equality $\min_{X \in \mathrm{Spect}_n} B \bullet X = \lambda_{\min}(B)$ follows by applying this with $C := -B$. For the second part, we just use the formula $B \bullet \tilde{X} = B \bullet \beta \beta^T = \beta^T B \beta$, easy to check and also mentioned in the proof of Theorem 4.8.1. $\qquad \square$

We will tackle the problem of computing δ-approximate smallest eigenvectors in Sect. 5.7. Fortunately, this amounts to approximating the largest eigenvalue of a suitable positive semidefinite matrix – a standard task in numerical mathematics.

In the Frank–Wolfe algorithm, Lemma 5.5.3 allows us to use a rank-1 matrix S_k of the form $S_k = \beta_k \beta_k^T$ to perform the update in Step 2.2. The matrix X_k will then be a sum of k rank-1 matrices, and as such it has rank at most k; see Exercise 5.4. This is relevant in applications where a sublinear number of iterations is performed in Algorithm 5.5.2. In this case, the algorithm is said to compute a *low-rank approximation*.

The Curvature Constant

Figure 5.1 indicates that the decrease of the f-value made during one iteration of Step 2 depends on the "shape" of f – the flatter, the better. Our goal is to relate the progress in the value of $f(X)$ to the progress in the value of $\nabla f(X_k) \bullet X$, but for this it is important to know how quickly f "rises above" its linearization as we move from X_k to X_{k+1}.

It is precisely the second derivative of f (geometrically corresponding to the curvature of the graph of f) that captures this. Thus, we start the analysis by assigning a "curvature constant" to f. This can be thought of as the maximum curvature of the graph of f over the spectahedron.

5.5.4 Definition. *Let $\nabla^2 f(X)$ be the Hessian of f at X, i.e., the matrix*

$$\nabla^2 f(X) := \left(\frac{\partial^2 \mathbf{f}}{\partial x_i \partial x_j}(\mathbf{X}) \right)_{1 \le i, j \le n^2} \in \mathrm{SYM}_{n^2}$$

of second partial derivatives of f at X. (See the paragraph before Lemma 5.4.2 for the boldface notation.)

The *curvature constant* $\Gamma(f)$ of f is defined as

$$\Gamma(f) = \max_{X \in \text{Spect}_n} \lambda_{\max}(\nabla^2 f(X)),$$

where $\lambda_{\max}(M)$ *denotes the maximum eigenvalue of a (symmetric square) matrix M.*

We note that the maximum exists by continuity of the second partial derivatives, continuity of the largest eigenvalue, and compactness of the spectahedron. We also observe that all Hessians under consideration are positive semidefinite, which is a consequence of the convexity of f [PSU88, Sect. 2.3]. In particular, $\Gamma(f) \geq 0$.

Let us see that we can indeed use the curvature constant to bound the deviation of f from its linearization. Namely, in the setting of the algorithm, we have the following estimate.

5.5.5 Lemma. *Let $f, X_k, S_k, \alpha_k, X_{k+1}$ be as in Algorithm 5.5.2, and set $Z := \alpha_k(S_k - X_k) = X_{k+1} - X_k$. Then*

$$f(X_{k+1}) \leq f(X_k) + \nabla f(X_k) \bullet Z + \alpha_k^2 \Gamma(f).$$

In Fig. 5.1, the lemma claims that at the point X_{k+1}, the vertical distance between f and the linearization is at most $\alpha_k^2 \Gamma(f)$. (We recall that α_k is the step size parameter in the k-th iteration of Algorithm 5.5.2.)

Proof. We use Taylor's formula [PSU88, Theorem 1.2.4], according to which there exists a convex combination X' of X_k and X_{k+1} such that

$$f(X_{k+1}) = f(X_k) + \nabla f(X_k) \bullet Z + \frac{1}{2} \mathbf{Z}^T \nabla^2 f(X') \mathbf{Z}.$$

We also have

$$\|\mathbf{Z}\|^2 = \|Z\|_F^2 = \alpha_k^2 \|S_k - X_k\|_F^2 \leq 2\alpha_k^2$$

(since Spect_n has *diameter* $\sqrt{2}$; see Exercise 5.6). Now we use that

$$\lambda_{\max}(A) = \max_{\|\mathbf{u}\|=1} \mathbf{u}^T A \mathbf{u}$$

for every matrix A (Theorem 4.8.1) to conclude that

$$\mathbf{Z}^T \nabla^2 f(X') \mathbf{Z} \leq \|\mathbf{Z}\|^2 \lambda_{\max}(\nabla^2 f(X')) \leq 2\alpha_k^2 \Gamma(f),$$

due to $X' \in \text{Spect}_n$. The statement follows. $\qquad\square$

The Analysis

We will prove the following result on the convergence rate of the Frank–Wolfe algorithm.

5.5.6 Proposition. *Let X^* be an optimal solution of the convex optimization problem $\min_{X \in \text{Spect}_n} f(X)$, and let X_k be the k-th iterate computed by Algorithm 5.5.2. Then, for all $k = 1, 2, \ldots,$*

$$f(X_k) \leq f(X^*) + \frac{\varepsilon}{4} + \frac{4\Gamma(f)}{k}.$$

Proof. The main ingredient of the proof is the inequality

$$f(X_{k+1}) \leq f(X_k) + \nabla f(X_k) \bullet \alpha_k(S_k - X_k) + \alpha_k^2 \Gamma(f) \qquad (5.9)$$

that we derived in Lemma 5.5.5. Then we need the fact that S_k is approximately optimal for the linearized problem; namely,

$$\nabla f(X_k) \bullet S_k \leq \min_{X \in \text{Spect}_n} \nabla f(X_k) \bullet X + \frac{\varepsilon}{4} \leq \nabla f(X_k) \bullet X^* + \frac{\varepsilon}{4}. \qquad (5.10)$$

The last ingredient is the convexity of f. Geometrically, this means that the graph of f is above the graph of any of its linearizations. Here we use this for the linearization $\ell(X)$ at X_k:

$$f(X_k) + \nabla f(X_k) \bullet (X^* - X_k) \leq f(X^*). \qquad (5.11)$$

Starting from (5.9) and plugging in (5.10) and (5.11), we arrive at

$$f(X_{k+1}) \leq (1 - \alpha_k) f(X_k) + \alpha_k \left(f(X^*) + \frac{\varepsilon}{4} \right) + \alpha_k^2 \Gamma(f). \qquad (5.12)$$

From this, we derive the statement of the proposition by induction. For brevity, let us write $\omega := f(X^*) + \frac{\varepsilon}{4}$.

For $k = 1$, we have $\alpha_k = 1$, and (5.12) yields $f(X_2) \leq \omega + \Gamma(f)$ as required. Now let us assume that the statement holds for some $k \geq 2$. With $\alpha_k = 2/k$, (5.12) gives

$$f(X_{k+1}) \leq (1 - \alpha_k) \left(\omega + \frac{4\Gamma(f)}{k} \right) + \alpha_k \omega + \alpha_k^2 \Gamma(f)$$

$$= \omega + \left(1 - \frac{2}{k} \right) \frac{4\Gamma(f)}{k} + \frac{4\Gamma(f)}{k^2} = \omega + \frac{k-1}{k^2} \cdot 4\Gamma(f) \leq \omega + \frac{4\Gamma(f)}{k+1}$$

by a simple calculation. $\square$

The proposition just proved gives us a bound for the required number of iterations in the Frank–Wolfe algorithm.

5.5.7 Corollary. *For*
$$k \geq 16\Gamma(f)/\varepsilon,$$
the iterate X_k in Algorithm 5.5.2 satisfies $f(X_k) \leq f(X^*) + \varepsilon/2$. ☐

We remark that up to a constant, this bound on the number of iterations cannot be improved in the worst case; see Exercise 5.8.

5.6 Back to the Semidefinite Feasibility Problem

In order to apply Corollary 5.5.7 with $f = f_{\text{pen}}$, we need an upper bound for $\Gamma(f_{\text{pen}})$.

5.6.1 Lemma. $\Gamma(f_{\text{pen}}) \leq K = \dfrac{2 \log m}{\varepsilon}$.

Proof. According to Definition 5.5.4, we have to bound the largest eigenvalue of every Hessian $\nabla^2 f_{\text{pen}}(X), X \in \text{Spect}_n$.

We know (Exercise 4.8) that the sum of eigenvalues of a matrix equals its trace. Thus, for positive semidefinite matrices, such as the Hessians we consider, the largest eigenvalue is bounded by the trace. Furthermore, it is easy to prove (see Exercise 5.7) that

$$\text{Tr}(\nabla^2 f_{\text{pen}}(X)) \leq K \sum_{i=1}^{m} y_i(X) \underbrace{\|A_i\|_F^2}_{=1}, \tag{5.13}$$

where the $y_i(X)$ are nonnegative coefficients summing up to one. ☐

We also need to show how we can approximately solve the linearized problem in Step 2.1 of the Frank–Wolfe algorithm. In Sect. 5.7 below, we will prove the following.

5.6.2 Proposition. *Given $\nabla f(X_k)$, a matrix $S_k \in \text{Spect}_n$ satisfying*

$$\nabla f_{\text{pen}}(X_k) \bullet S_k \leq \min_{X \in \text{Spect}_n} \nabla f_{\text{pen}}(X_k) \bullet X + \frac{\varepsilon}{4}$$

can be computed with

$$O\left(\frac{N}{\varepsilon} \log n\right)$$

arithmetic operations, with high probability. Moreover, we obtain S_k of the form $\text{ß}_k \text{ß}_k^T$, with ß_k a unit vector.

Now we have all ingredients together to derive our main theorem from
Sect. 5.3, which bounds the number of arithmetic operations necessary to
solve the feasibility problem.

Proof of Theorem 5.3.2. We set

$$t := \frac{16K}{\varepsilon} = O\left(\frac{\log m}{\varepsilon^2}\right),$$

and we perform t iterations of the Frank–Wolfe Algorithm 5.5.2 with the
penalty function f_{pen}. According to Corollary 5.5.7 and Lemma 5.4.1, X_t is
then either the desired ε-approximate solution of the feasibility problem, or
a certificate that there is no feasible solution.

It remains to count the arithmetic operations. Here is what we do in the
k-th iteration:

- We compute the gradient $\nabla f_{\mathrm{pen}}(X_k)$, which can be done with $O(N)$ arith-
  metic operations; see Lemma 5.4.2.
- We approximately solve the linearized problem $\min_{X \in \mathrm{Spect}_n} \nabla f_{\mathrm{pen}}(X_k) \bullet X$.
  This can be done with $O(N(\log n)/\varepsilon)$ arithmetic operations by Proposi-
  tion 5.6.2.
- We compute X_{k+1}. By Lemma 5.4.2 again, the next iteration does not
  require the matrix X_{k+1} itself but merely the scalar products $A_i \bullet X_{k+1}$.
  If we already have $A_i \bullet X_k$, we compute

$$A_i \bullet X_{k+1} = (1 - \alpha_k)(A_i \bullet X_k) + \alpha_k(A_i \bullet S_k), \quad i = 1, 2, \ldots, m,$$

  and the latter scalar products are obtained with $O(N)$ arithmetic operations for all i, due to $S_k = \text{\ss}_k\text{\ss}_k^T$.

The bound claimed in the theorem follows by combining these bounds.
To get the high probability result for the whole algorithm, the number $t = O((\log m)/\varepsilon^2)$ of iterations has to be bounded by a polynomial in n. $\quad\square$

5.7 From the Linearized Problem to the Largest Eigenvalue

By Lemma 5.5.3, the matrix S_k in iteration k of the Frank–Wolfe algorithm
can be chosen of the form $S_k = \text{\ss}_k\text{\ss}_k^T$, where the vector $\text{\ss}_k$ is an approximate
eigenvector belonging to the smallest eigenvalue of the symmetric matrix $B = \nabla f(X_k)$. The next lemma allows us to further transform this computational
task to computing an approximate eigenvector for the *largest* eigenvalue of a
positive definite matrix.

5.7.1 Lemma. *Let $B \in \mathrm{SYM}_n$, $B \neq 0$, $\|B\|_F \leq 1$, and let us set $Q := 2I_n - B$. Then the following statements hold.*

(i) *Q is positive definite.*

(ii) *$\lambda_{\max}(Q) = 2 - \lambda_{\min}(B) \in [1, 3]$.*

(iii) *If $\mathbf{u} \in \mathbb{R}^n$, $\|\mathbf{u}\| = 1$, and $\mathbf{u}^T Q \mathbf{u} \geq \lambda_{\max}(Q) - \delta$, then $\mathbf{u}^T B \mathbf{u} \leq \lambda_{\min}(B) + \delta$.*

Proof. We observe that all eigenvalues of B (as well as those of $-B$) lie in the interval $[-1, 1]$. Indeed, for an eigenvalue λ of B with a unit eigenvector $\mathbf{x}$, we have

$$|\lambda| = |\mathbf{x}^T B \mathbf{x}| = |B \bullet \mathbf{x}\mathbf{x}^T| \leq \|B\|_F \|\mathbf{x}\mathbf{x}^T\|_F = \|B\|_F \leq 1,$$

by the Cauchy–Schwarz inequality. It follows that all eigenvalues of Q are in the interval $[1, 3]$. In particular, Q is positive definite, and this shows (i). Parts (ii) and (iii) are simple calculations. $\qquad\square$

As was announced in the introduction, we will employ the power method for computing an approximate eigenvector. In the next section, we will establish the following theorem.

5.7.2 Theorem. *Let Q be an $n \times n$ positive definite matrix with at most N nonzero entries, such that $\lambda_{\max}(Q) \in [1, 3]$. Then, given $\delta \in (0, \frac{1}{2}]$, we can compute a unit δ-approximate eigenvector $\mathbf{u}$ for the largest eigenvalue of Q, i.e., a $\mathbf{u}$ such that*

$$\|\mathbf{u}\| = 1, \quad \mathbf{u}^T Q \mathbf{u} \geq \lambda_{\max}(Q) - \delta.$$

With high probability, the number of arithmetic operations can be bounded by

$$O\left(\frac{N}{\delta} \log n\right).$$

Now we are ready to provide the oracle for the Frank–Wolfe algorithm promised in Proposition 5.6.2.

Proof of Proposition 5.6.2. In the setting of the proposition, we have $B = \nabla f_{\mathrm{pen}}(X_k)$, and by Lemma 5.4.2, which describes the structure of ∇f_{pen}, we know that B has at most N nonzero entries and satisfies $\|B\|_F \leq \max_i \|A_i\|_F = 1$. Thus, we can use Lemma 5.7.1, set up the matrix Q, and compute an approximate eigenvector of it according to Theorem 5.7.2, with $\delta := \varepsilon/4$. It is easy to check, using Lemma 5.7.1(iii), that the resulting approximate eigenvector can be used as $ß_k$ in the proposition. $\qquad\square$

5.8 The Power Method

The power method is a well-known numerical method for finding a dominant eigenvalue of a square matrix Q, plus a corresponding eigenvector. (An eigenvalue is *dominant* if it has the largest absolute value among all eigenvalues of Q.)

Given a start vector $\mathbf{u}_0 \in \mathbb{R}^n$ (whose choice will be discussed starting from page 95 below), the power method computes a sequence $\mathbf{u}_1, \mathbf{u}_2, \ldots$ of unit vectors as follows.

5.8.1 Algorithm (The power method).

1. Let $\mathbf{u}_0 \in \mathbb{R}^n$, $\mathbf{u}_0 \neq \mathbf{0}$.
2. For $k = 0, 1, 2 \ldots$ do:

   2.1 $\mathbf{v}_k := Q\mathbf{u}_k$,
   2.2 $\varrho_k := \mathbf{u}_k^T \mathbf{v}_k$,
   2.3 $\mathbf{u}_{k+1} := \mathbf{v}_k / \|\mathbf{v}_k\| \in S^{n-1}$.

It is easy to show by induction that

$$\mathbf{u}_k = \frac{Q^k \mathbf{u}_0}{\|Q^k \mathbf{u}_0\|}, \quad k = 1, 2, \ldots,$$

and this explains the name "power method."

Under suitable conditions, ϱ_k converges to a dominant eigenvalue of Q as $k \to \infty$, and $\mathbf{u}_k$ to a corresponding eigenvector.[3]

One of the conditions is that the initial vector $\mathbf{u}_0$ is not orthogonal (or very close to orthogonal) to the corresponding eigenvector. If we choose $\mathbf{u}_0$ at random, for example, then it is very unlikely to be almost orthogonal to a fixed vector, and so we will tacitly assume in the following discussion that this condition is satisfied (we will return to it on page 95).

The convergence of the power method is rather fast, and easy to prove, *provided that* Q has only a single dominant eigenvalue λ_1, and moreover, $|\lambda_i/\lambda_1| \leq \alpha$ holds for all the other eigenvalues λ_i, $i = 2, 3, \ldots, n$, where $\alpha < 1$ is a constant. Then the error $|\varrho_k - \lambda_1|$ decreases geometrically, like α^k; see Exercise 5.10.[4]

---

[3] In principle, it may happen that one of the $\mathbf{v}_k$ equals $\mathbf{0}$, and then $\mathbf{u}_{k+1}$ is not even well defined. However, if Q has full rank (in particular, if it is positive definite), then $\mathbf{u}_0 \neq \mathbf{0}$ implies $\mathbf{u}_k \neq \mathbf{0}$ for all k, and we need not worry about this issue.

[4] The exercise makes the claim about convergence for Q positive definite, but it actually holds for arbitrary Q, possibly with complex entries and not necessarily symmetric – the assumption of a single dominant eigenvalue is sufficient. In this context, one can observe that if Q is real and has a single dominant eigenvalue λ_1, then λ_1 has to be real, for example because complex eigenvalues of a real matrix come in conjugate pairs.

However, the method becomes much slower if there are several dominant eigenvalues, and in some cases, it may even fail to converge.

For us, it suffices to apply the power method only to *positive semidefinite* matrices Q. Then a considerably more sophisticated analysis shows that the power method converges for *all* such Q, but the convergence may be much slower; the error after k iterations may decrease only like k^{-1}.

From now on, we will thus assume that Q is as in Theorem 5.7.2, i.e., positive definite and with $\lambda_{\max}(Q) \in [1,3]$. Let the eigenvalues of Q be $\lambda_{\max}(Q) = \lambda_1 \geq \lambda_2 \geq \cdots \geq \lambda_n > 0$, and let $\mathbf{x}_1, \ldots, \mathbf{x}_n$ be the corresponding unit eigenvectors.

Our convergence analysis for ϱ_k starts with the following lemma.

5.8.2 Lemma. *Let* $w_1, \ldots, w_n$ *be the unique coefficients such that*

$$\mathbf{u}_0 = \sum_{i=1}^{n} w_i \mathbf{x}_i. \tag{5.14}$$

Then

$$\varrho_k = \frac{\sum_{i=1}^{n} w_i^2 \lambda_i^{2k+1}}{\sum_{i=1}^{n} w_i^2 \lambda_i^{2k}}, \quad k = 1, 2, \ldots. \tag{5.15}$$

The simple proof (that works for any square matrix Q) is left as Exercise 5.9. Suppose for a moment that $\lambda_1 > \lambda_2$ and that $w_1 \neq 0$ (meaning that $\mathbf{u}_0$ is not orthogonal to $\mathbf{x}_1$). Then for large k, the terms $w_1^2 \lambda_1^{2k+1}$ and $w_1^2 \lambda_1^{2k}$ dominate numerator and denominator of (5.15), so the fraction converges to λ_1.

Here we are interested in the *rate* of convergence, and we also want to deal with the case $\lambda_1 = \lambda_2$. Not surprisingly, the convergence rate depends on "how orthogonal" $\mathbf{u}_0$ is to $\mathbf{x}_1$. We measure this by the quantity

$$\tau = \tau(\mathbf{u}_0) = \frac{w_2^2 + \cdots + w_n^2}{w_1^2} \in [0, \infty],$$

where $w_1, \ldots, w_n$ are as in Lemma 5.8.2. The larger $\tau(\mathbf{u}_0)$, the worse the power method converges. (It is easy to show that $\tau = \tan^2 \alpha$, where α is the angle between $\mathbf{u}_0$ and $\mathbf{x}_1$.)

Now we are ready to state the main result of this section.

5.8.3 Proposition. *Suppose that* $w_1 \neq 0$, *and let* $\beta \in (0, 1/2]$ *be a real number. If*

$$2k + 1 > \frac{\max(1, \ln \tau)}{|\ln(1 - \beta)|},$$

then the number ϱ_k *in Algorithm 5.8.1 satisfies*

$$\lambda_1 \geq \varrho_k \geq (1 - \beta)\lambda_1.$$

Proof. We follow [OSV79]. Since the claimed inequalities $\lambda_1 \geq \varrho_k \geq (1-\beta)\lambda_1$ are easily seen to be invariant under scaling Q, we may assume w.l.o.g. that $\lambda_1 = 1$. Then (5.15) becomes

$$\varrho_k = \frac{w_1^2 + \sum_{i=2}^n w_i^2 \lambda_i^{2k+1}}{w_1^2 + \sum_{i=2}^n w_i^2 \lambda_i^{2k}} =: \frac{g_{2k+1}}{g_{2k}}, \tag{5.16}$$

with $w_1^2 > 0$ by assumption. The inequality $1 \geq \varrho_k$ is immediate from $\lambda_2, \ldots, \lambda_n \leq 1$. The claimed lower bound follows from the next two lemmas. □

5.8.4 Lemma. *The value ϱ_k as defined in (5.16) is bounded from below by the unique root $\tilde{\varrho}_k \in (0,1)$ of the polynomial*

$$p_k(x) := \frac{c_k \tau}{2k+1} x^{2k+1} + x - 1, \text{ where } c_k := \left(\frac{2k}{2k+1}\right)^{2k}. \tag{5.17}$$

5.8.5 Lemma. *Let $\tilde{\varrho}_k$ be the unique root of the polynomial $p_k(x)$ (as in Lemma 5.8.4) lying in $(0,1)$. If*

$$2k+1 \geq \frac{\max(1, \ln \tau)}{|\ln(1-\beta)|},$$

then $\tilde{\varrho}_k \geq 1 - \beta$.

Proof of Lemma 5.8.4. We determine the distribution of $(\lambda_2, \ldots, \lambda_n) \in [0,1]^{n-1}$ for which the expression (5.16) (considered as a function of $\lambda_2, \ldots, \lambda_n$) is minimized. There must be a minimum over the compact set $[0,1]^{n-1}$ by the continuity of ϱ_k.

Fixing the values of λ_j, $j \neq i$, arbitrarily, we have

$$\varrho_k = \frac{W + w_i^2 \lambda_i^{2k+1}}{W' + w_i^2 \lambda_i^{2k}}, \quad W \leq W'.$$

Elementary calculations then show that ϱ_k is minimized for $\lambda_i \in (0,1)$. This means that every minimum of ϱ_k is an interior point of $[0,1]^{n-1}$, at which necessarily $\nabla \varrho_k = \mathbf{0}$.

We next prove that there are unique positive values $\tilde{\lambda}_2, \ldots, \tilde{\lambda}_n$ for which the gradient $\nabla \varrho_k$ vanishes. This means that ϱ_k has a unique minimum $\tilde{\varrho}_k$ at $\tilde{\lambda}_2, \ldots, \tilde{\lambda}_n$.

For this, we first compute the partial derivatives and get

$$\frac{\partial \varrho_k}{\partial \lambda_i} = \frac{(2k+1)w_i^2 \lambda_i^{2k} g_{2k} - 2k w_i^2 \lambda_i^{2k-1} g_{2k+1}}{g_{2k}^2}, \quad i = 2, \ldots, n.$$

The unique positive zero of $\partial \varrho_k / \partial \lambda_i$ is

$$\lambda_i = \frac{2k}{2k+1} \frac{g_{2k}}{g_{2k+1}} = \frac{2k}{2k+1} \varrho_k < 1, \quad i = 2, \ldots, n.$$

To find the values $\tilde{\lambda}_i$ and establish their uniqueness, we need to solve the polynomial equation

$$\varrho_k = \frac{w_1^2 + \left(\frac{2k}{2k+1}\varrho_k\right)^{2k+1} \sum_{i=2}^n w_i^2}{w_1^2 + \left(\frac{2k}{2k+1}\varrho_k\right)^{2k} \sum_{i=2}^n w_i^2} = \frac{1 + \left(\frac{2k}{2k+1}\varrho_k\right)^{2k+1} \tau}{1 + \left(\frac{2k}{2k+1}\varrho_k\right)^{2k} \tau}$$

and argue that it has a unique solution $\tilde{\varrho}_k$. Multiplying with the denominator, this equation simplifies to (5.17). Since we have $p_k'(x) = c_k \tau x^{2k} + 1$, the polynomial $p_k(x)$ is monotone increasing for $0 \leq x \leq 1$. Since we have $p_k(0) = -1$ and $p_k(1) = c_k \tau/(2k+1) > 0$, there is indeed a unique zero $x = \tilde{\varrho}_k \in (0,1)$, which corresponds to the minimum of ϱ_k over $[0,1]^{n-1}$. The lemma is proved. $\square$

Proof of Lemma 5.8.5. Let us set $\alpha := 1 - \beta \in [1/2, 1)$. As shown in the proof of Lemma 5.8.4, the polynomial $p_k(x)$ defined in (5.17) is monotone increasing. Hence, if $p_k(\alpha) < 0$, we have $\tilde{\varrho}_k \geq \alpha$. With k as in the statement of the lemma,

$$\alpha^{2k+1} \leq \alpha^{-\max(1, \ln \tau)/\ln \alpha} = e^{-\max(1, \ln \tau)} = \min\left(\frac{1}{e}, \frac{1}{\tau}\right). \qquad (5.18)$$

Moreover, for $1/2 \leq \alpha < 1$, we have $(\alpha - 1) \leq (\ln \alpha)/2$, and so

$$p_k(\alpha) = \frac{c_k \tau}{2k+1} \alpha^{2k+1} + \alpha - 1 \leq \frac{c_k}{2k+1} + \frac{\ln \alpha}{2}.$$

Taking logarithms in (5.18), we obtain $\ln \alpha \leq -1/(2k+1)$. Then

$$p_k(\alpha) \leq \frac{c_k - \frac{1}{2}}{2k+1} \leq \frac{c_1 - \frac{1}{2}}{2k+1} < 0,$$

since $c_1 = \frac{4}{9} < \frac{1}{2}$, and the lemma is proved. $\square$

The Start Vector

In order for the power method to converge (quickly), we need a start vector $\mathbf{u}_0$ for which $\ln \tau$ is not too large, where

$$\tau = \tau(\mathbf{u}_0) = \frac{w_2^2 + \cdots + w_n^2}{w_1^2}.$$

The following lemma shows that a random $\mathbf{u}_0$ works with high probability.

5.8.6 Lemma. *Let $\mathbf{u}_0 \in \mathbb{R}^n$ be a vector whose n components are independent $N(0,1)$ variables (standard Gaussians). For every $c > 0$,*

$$\mathrm{Prob}[\ln \tau(\mathbf{u}_0) \leq (3c+1)\ln n] \geq 1 - \frac{2}{n^c}.$$

Proof. For a standard Gaussian w, $\mathbf{E}\left[w^2\right] = 1$, hence $\mathbf{E}\left[w_2^2 + \cdots + w_n^2\right] = n - 1$. By the Markov inequality,

$$\mathrm{Prob}\left[w_2^2 + \cdots + w_n^2 \geq n^{c+1}\right] < \frac{1}{n^c}. \tag{5.19}$$

We also have

$$\mathrm{Prob}\left[|w_1| \leq \frac{1}{n^c}\right] = \frac{1}{\sqrt{2\pi}} \int_{-1/n^c}^{1/n^c} e^{-x^2/2}\, dx \leq \frac{1}{\sqrt{2\pi}} \int_{-1/n^c}^{1/n^c} dx \leq \frac{1}{n^c}. \tag{5.20}$$

With (5.19) and (5.20), the union bound yields

$$
\begin{aligned}
\mathrm{Prob}\left[\tau(\mathbf{u}_0) \leq n^{3c+1}\right] &\geq \mathrm{Prob}\left[w_2^2 + \cdots + w_n^2 \leq n^{c+1} \text{ and } w_1^2 \geq \frac{1}{n^{2c}}\right] \\
&\geq 1 - \frac{2}{n^c},
\end{aligned}
$$

and the lemma follows. $\square$

Putting It Together

We can now prove Theorem 5.7.2, stating that an eigenvector belonging to the largest eigenvalue of a positive definite matrix Q can be computed efficiently.

Proof of Theorem 5.7.2. We set $\beta := \delta/\lambda_{\max}(Q)$. We have $\delta/3 \leq \beta \leq \delta \leq 1/2$ (since we assume $\lambda_{\max}(Q) \in [1,3]$). According to Proposition 5.8.3, the value ϱ_k in Algorithm 5.8.1 satisfies

$$\mathbf{u}_k^T Q \mathbf{u}_k = \varrho_k \geq (1-\beta)\lambda_{\max}(Q) = \lambda_{\max}(Q) - \delta$$

for $2k+1 > \max(1, \ln \tau(\mathbf{u}_0))/|\ln(1-\beta)|$, so $\mathbf{u}_k$ is the desired δ-approximate eigenvector of Q.

For $\beta \in (0,1)$, we have $|\ln(1-\beta)| \geq \beta \geq \delta/3$. Hence we need to perform

$$k = O\left(\frac{1}{\delta}\max(1, \log \tau(\mathbf{u}_0))\right)$$

iterations of the power method, where every iteration takes time $O(N)$. The statement of the theorem follows, since with probability at least $1 - 2/n^c$, we have $\log \tau(\mathbf{u}_0) = O(\log n)$ by Lemma 5.8.6. $\qquad\square$

The bound of Theorem 5.7.2 on the required number of iterations cannot be improved in the worst case [KW92]. But if the two largest eigenvalues are well-separated, the bound drops from $O(\frac{\log n}{\delta})$ to $O(\log \frac{n}{\delta})$. This is Exercise 5.10.

Exercises

5.1 Prove the following variant of Theorem 5.2.2. *Let* $\mathsf{SDP}(G)$ *be the maximum value of the semidefinite relaxation of the* MAXCUT *problem, for a given graph* $G = (V, E)$ *with* n *vertices. For every constant* $\varepsilon \in (0, \frac{1}{2}]$, *we can compute a matrix* $\tilde{X} \succeq 0$ *such that*

(i) $\tilde{x}_{ii} = 1$ for $i = 1, 2, \ldots, n$
(ii) $\sum_{\{i,j\} \in E} \frac{1 - \tilde{x}_{ij}}{2} \geq \mathsf{SDP}(G) - \varepsilon |E|$

With high probability, The algorithm requires

$$O\left((n^4 + n^3 |E|) \log^3 n \right)$$

arithmetic operations, with a hidden cubic dependence on $1/\varepsilon$.

Hint: Start from the matrix $\tilde{X}'$ as guaranteed by Theorem 5.2.2 and set $\tilde{X} = D\tilde{X}'D$ for a diagonal matrix $D \succ 0$ chosen so that $\tilde{x}_{ii} = 1$ for all i.

5.2 Let $A_1, A_2, \ldots, A_m \in \mathrm{SYM}_n$ be symmetric matrices, $\mathbf{b} \in \mathbb{R}^m$. Show that the function $f_{\mathrm{pen}} \colon \mathbb{R}^{n \times n} \to \mathbb{R}$ defined by

$$f_{\mathrm{pen}}(X) := \frac{1}{K} \log \left(\sum_{i=1}^{m} e^{K(A_i \bullet X - b_i)} \right)$$

is convex over $\mathbb{R}^{n \times n}$, for all real numbers $K > 0$.

Hint: You may proceed in the following steps.

(i) Prove that the function $f \colon \mathbb{R}^m \to \mathbb{R}$ defined by $f(\mathbf{x}) = \frac{1}{K} \log \left(\sum_{i=1}^{m} e^{Kx_i} \right)$ is convex. For this, it is sufficient that all Hessians are positive semidefinite [PSU88, Theorem 2.3.7].
(ii) Consider the linear functions $g_i \colon \mathbb{R}^{n \times n} \to \mathbb{R}$, where $g_i(X) = A_i \bullet X - b_i$. Derive the convexity of $f_{\mathrm{pen}}(X) = f(g_1(X), \ldots, g_m(X))$ from the convexity of f and the linearity of the g_i.

5.3 Let $K \in \mathbb{R}$, $K > 0$, and consider the function $\Phi_K \colon \mathbb{R}^m \to \mathbb{R}$ defined by

$$\Phi_K(\mathbf{y}) = \frac{1}{K} \log\left(\sum_{i=1}^{m} e^{K y_i} \right).$$

Prove that

$$\max{}_{i=1}^{m} y_i \leq \Phi_K(\mathbf{y}) \leq \max{}_{i=1}^{m} y_i + \frac{\log m}{K}, \qquad \text{for all } \mathbf{y} \in \mathbb{R}^m.$$

5.4 Let $M \in \mathbb{R}^{n \times n}$, $\mathbf{v} \in \mathbb{R}^n$. Prove that

$$\mathrm{rank}(M + \mathbf{v}\mathbf{v}^T) \leq \mathrm{rank}(M) + 1.$$

5.5 Prove that $\|X\|_F \leq 1$ for every $X \in \mathrm{Spect}_n = \{X \in \mathrm{SYM}_n : \mathrm{Tr}(X) = 1,\ X \succeq 0\}$.

5.6 Prove that $\|X - Y\|_F \leq \sqrt{2}$ for all $X, Y \in \mathrm{Spect}_n$, and that this bound is tight.

5.7 Prove inequality (5.13):

$$\mathrm{Tr}(\nabla^2 f_{\mathrm{pen}}(X)) \leq K = \frac{2 \log m}{\varepsilon}.$$

5.8 Show that the analysis of Hazan's algorithm cannot be improved in the worst case. For this, prove that the bound of Corollary 5.5.7 on the number of iterations is asymptotically tight for the convex function $f(X) = \|X\|_F^2$ if $k \leq n$ and $\varepsilon = 2/n$.

Hint: Prove that the function $f(X) = \|X\|_F^2$ has curvature constant $\Gamma(f) = 2$. Moreover, for $1 \leq k \leq n$,

$$\min_{\substack{X \in \mathrm{Spect}_n \\ \mathrm{rank}(X)=k}} f(X) = \frac{1}{k}.$$

5.9 Prove Lemma 5.8.2 that provides a formula for ϱ_k in Algorithm 5.8.1.

5.10 Prove the following statement. *Suppose that we apply the power method with a random start vector to a $Q \succ 0$ for which $\lambda_2/\lambda_1 \leq \alpha < 1$, where λ_1, λ_2 are the two largest eigenvalues of Q, and α is a constant. Then for every $\beta \in (0, 1]$, we have $\varrho_k \geq (1 - \beta)\lambda_1$ for*

$$k = \Theta\left(\log \frac{n}{\beta} \right),$$

with high probability.

Chapter 6
An Interior-Point Algorithm for Semidefinite Programming

In this chapter we consider another method for approximately solving semidefinite programs, a *primal-dual central path algorithm*. This algorithm belongs to the family of *interior-point methods*, whose scope reaches far beyond semidefinite programming.

The primal-dual central path algorithm exhibits some high-level similarities to Hazan's algorithm discussed in the previous chapter. Both work in the real RAM model of computation, and both compute approximately optimal solutions through a sequence of improvement steps.

The major advantage of the interior-point approach considered here is a vastly better dependence of the running time on the desired approximation error ε. The version of Hazan's algorithm that we analyzed in detail has runtime proportional to $\frac{\log(1/\varepsilon)}{\varepsilon^3}$. In contrast, the runtime of the primal-dual central path algorithm is proportional only to $\log(1/\varepsilon)$. The price to pay for this efficiency is a mildly more complicated improvement step, and a substantially more complicated analysis, which we will not do in detail.

What we will present in full detail is the theoretical foundation of the primal-dual central path algorithm, namely the existence and uniqueness of the central path under suitable conditions. We also provide a detailed description of the algorithm's main step – following the central path – but its analysis will be only sketched. Also, we will not discuss "phase one" of the algorithm, whose task is to find a feasible starting point close to the central path.

Interior-point methods for semidefinite programming have been pioneered by Nesterov and Nemirovski [NN90, NN94] as well as Alizadeh [Ali95]. The work of Nesterov and Nemirovski develops an interior-point method for general convex programming problems, based on the ingenious concept of *self-concordant barrier functions*. The semidefinite programming case is then handled by exhibiting a suitable barrier function for this case.

Alizadeh, in contrast, starts from a specific interior-point method for linear programming and extends it to the semidefinite case. On the way tools are

B. Gärtner and J. Matoušek, *Approximation Algorithms and Semidefinite Programming*, DOI 10.1007/978-3-642-22015-9_6,
© Springer-Verlag Berlin Heidelberg 2012

developed that allows a similar extension of many other interior-point method for linear programming.

The algorithm that we present in more detail in this chapter is the *short-step method* of Kojima et al. [KSH97] and Monteiro [Mon97]. For the central path material, we stay close to Laurent and Rendl [LR05]. Throughout this chapter, we consider semidefinite programs in equational form:

$$
\begin{aligned}
\text{Maximize} \quad & C \bullet X \\
\text{subject to} \quad & A_i \bullet X = b_i, \quad i = 1, 2, \ldots, m, \\
& X \succeq 0.
\end{aligned}
\tag{6.1}
$$

As usual, we assume w.l.o.g. that C and the A_i are symmetric matrices.

6.1 The Idea of the Central Path

The main idea behind all central path interior-point methods is to get rid of the "difficult" nonlinear constraint $X \succeq 0$ by modifying the objective function. Namely, we add a *barrier function* to it, so that the modified objective function tends to $-\infty$ as we approach the boundary of the set $\mathrm{PSD}_n = \{X \in \mathrm{SYM}_n : X \succeq 0\}$ from the interior. Then we can drop the constraints $X \succeq 0$, since they will be "non-binding" at optimality.

Here is a concrete realization of this idea. For a real number $\mu > 0$, we consider the auxiliary problem

$$
\begin{aligned}
\text{Maximize} \quad & f_\mu(X) := C \bullet X + \mu \ln \det X \\
\text{subject to} \quad & A_i \bullet X = b_i, \quad i = 1, 2, \ldots, m, \\
& X \succ 0,
\end{aligned}
\tag{6.2}
$$

where $X \succ 0$ means that X is positive definite (all eigenvalues are strictly positive). Since all matrices on the boundary of PSD_n have at least one eigenvalue equal to 0, they are singular and satisfy $\det X = 0$. Thus, $\mu \ln \det X$ is indeed a barrier function in the above sense.

We would like to claim that under suitable conditions, the auxiliary problem has a unique optimal solution $X^*(\mu)$ for every $\mu > 0$, and that $C \bullet X^*(\mu)$ converges to the optimum value of (6.1) as $\mu \to 0$. Obviously, we need to assume that there is a feasible $X \succ 0$, but additional conditions will be needed as well. The set $\{X^*(\mu) : \mu > 0\}$ is known as the *central path*, because the barrier term $\mu \ln \det X$ pushes $X^*(\mu)$ "towards the center" of the feasible region.

We will proceed in the following steps, always assuming that μ is strictly positive.

(i)   We show that *if* the auxiliary problem (6.2) has an optimal solution, then it has a *unique* optimal solution $X^*(\mu)$.

(ii) We derive necessary conditions for the existence of $X^*(\mu)$, in the form of a system of equations and inequalities that $X^*(\mu)$ has to satisfy. These conditions will also imply the desired convergence of $C \bullet X^*(\mu)$.

(iii) We prove that the necessary conditions derived in (ii) are also sufficient.

(iv) We show that the above necessary and sufficient conditions are satisfied under suitable assumptions on the semidefinite program. Thus, under these assumptions, the central path exists, and for μ small enough, $X^*(\mu)$ is an approximately optimal solution of the semidefinite program.

6.2 Uniqueness of Solution

Here is the first step in the above plan.

6.2.1 Lemma. *If f_μ attains a maximum over the feasible region of (6.2), then f_μ attains a unique maximum.*

Proof. This easily follows from the fact that f_μ is *strictly concave* over the interior of PSD_n (see the next lemma), meaning that for all $X, Y \succ 0$ with $X \neq Y$,

$$f_\mu((1-t)X + tY) > (1-t)f_\mu(X) + tf_\mu(Y), \quad 0 < t < 1.$$

Indeed, if the maximum were attained for two different matrices X^* and Y^*, then strict concavity would imply that $(X^* + Y^*)/2$ has even higher f_μ-value – a contradiction. We note that all considered matrices $(1-t)X + tY$ are positive definite as well by convexity of (the interior of) PSD_n.   □

6.2.2 Lemma. *The function $X \mapsto \ln \det X$ is strictly concave on the interior of PSD_n. (Since $C \bullet X$ is linear in X, this also implies strict concavity of f_μ for every $\mu > 0$.)*

Proof. For fixed matrices $X, Y \succ 0$, $X \neq Y$, we define

$$g(t) = \ln \det((1-t)X + tY).$$

If we can prove that g is strictly concave on the interval $[0, 1]$, we are done, since then

$$\underbrace{\ln \det((1-t)X + tY)}_{g((1-t)0+t1)} > \underbrace{(1-t)\ln \det X + t \ln \det Y}_{(1-t)g(0)+tg(1)}, \quad 0 < t < 1.$$

Because $X, Y \succ 0$, the function g is actually defined on some open interval containing $[0, 1]$. Moreover, g has derivatives of all orders.

Let us first recall how we can prove strict concavity of a *twice differentiable* function in one real variable over an open set U. A sufficient condition from analysis is that the second derivative is negative throughout U. For example, to prove that $f(x) = \ln(x)$ is strictly concave over the positive reals, we compute $f''(x) = -1/x^2 < 0$ for all $x > 0$.

Let us now write $g(t)$ as

$$g(t) = \ln \det(X + tZ), \quad Z = Y - X \in \mathrm{SYM}_n.$$

Since $X \succ 0$, there is a nonsingular matrix U such that $X = U^T U$, and we can write

$$X + tZ = U^T (I_n + t(U^{-1})^T Z U^{-1}) U.$$

Using $\det U = \det U^T$, together with multiplicativity of the determinant and the properties of logarithms, gives

$$
\begin{aligned}
\ln \det(X + tZ) &= 2\ln \det U + \ln \det(I_n + t(U^{-1})^T Z U^{-1}) \\
&= \ln \det X + \ln \det(I_n + t\tilde{Z}),
\end{aligned}
$$

with $\tilde{Z} := (U^{-1})^T Z U^{-1} \in \mathrm{SYM}_n$. Let $\tilde{Z}$ have eigenvalues $\lambda_1, \ldots, \lambda_n$. Then $I_n + t\tilde{Z}$ has eigenvalues $1 + t\lambda_1, \ldots, 1 + t\lambda_n$, and since the determinant of a symmetric matrix is the product of its eigenvalues (another fact that we get from diagonalization), we further have

$$g(t) = \ln \det(X + tZ) = \ln \det X + \sum_{i=1}^{n} \ln(1 + t\lambda_i).$$

This yields

$$g'(t) = \sum_{i=1}^{n} \frac{\lambda_i}{1 + t\lambda_i}, \quad g''(t) = -\sum_{i=1}^{n} \frac{\lambda_i^2}{(1 + t\lambda_i)^2}.$$

Since $X \neq Y$, we have $\tilde{Z} \neq 0$, so least one λ_i is nonzero, and we obtain $g''(t) < 0$. $\qquad\square$

6.3 Necessary Conditions for Optimality

According to the previous section we know that if the auxiliary problem has an optimal solution at all, then it has a unique optimal solution $X^*(\mu)$. Now we use the method of *Lagrange multipliers* from analysis to derive a system of equations and inequalities that $X^*(\mu)$ has to satisfy if it exists at all.

The Method of Lagrange Multipliers

We recall that this is a general method for finding a (local) maximum of $f(\mathbf{x})$ subject to m constraints $g_1(\mathbf{x}) = 0$, $g_2(\mathbf{x}) = 0,\ldots$, $g_m(\mathbf{x}) = 0$, where f and $g_1,\ldots,g_m$ are functions from $\mathbb{R}^n$ to $\mathbb{R}$. It can be seen as a generalization of the basic calculus trick for maximizing a univariate function by seeking a zero of its derivative. It introduces the following system of equations with unknowns $\mathbf{x} \in \mathbb{R}^n$ and $\mathbf{y} \in \mathbb{R}^m$ (the y_i are auxiliary variables called the *Lagrange multipliers*):

$$g_1(\mathbf{x}) = g_2(\mathbf{x}) = \cdots = g_m(\mathbf{x}) = 0 \quad \text{and} \quad \nabla f(\mathbf{x}) = \sum_{i=1}^{m} y_i \nabla g_i(\mathbf{x}). \qquad (6.3)$$

Here ∇ denotes the gradient (which, by convention, is a row vector):

$$\nabla f(\mathbf{x}) = \left(\frac{\partial f(\mathbf{x})}{\partial x_1}, \frac{\partial f(\mathbf{x})}{\partial x_2}, \ldots, \frac{\partial f(\mathbf{x})}{\partial x_n} \right).$$

That is, ∇f is a vector function from $\mathbb{R}^n$ to $\mathbb{R}^n$ whose i-th component is the partial derivative of f with respect to x_i. Thus, the equation $\nabla f(\mathbf{x}) = \sum_{i=1}^{m} y_i \nabla g_i(\mathbf{x})$ stipulates the equality of two n-component vectors.

6.3.1 Theorem (see e.g., [PSU88]). *Let f and the g_i be defined on a nonempty open subset U of $\mathbb{R}^n$ and have continuous first partial derivatives there. Let $\tilde{\mathbf{x}} \in U$ be a regular point, meaning that the vectors $\nabla g_i(\tilde{\mathbf{x}})$ are linearly independent.*

If $\tilde{\mathbf{x}}$ is a local maximum of $f(\mathbf{x})$ subject to $g_1(\mathbf{x}) = g_2(\mathbf{x}) = \cdots = g_m(\mathbf{x}) = 0$, then $\tilde{\mathbf{x}}$ satisfies (6.3); that is, there exists $\tilde{\mathbf{y}}$ such that $\tilde{\mathbf{x}}$ and $\tilde{\mathbf{y}}$ together fulfill (6.3).

If the constraint functions g_i are *linear* (as they will be in our application), the regularity requirement on $\tilde{\mathbf{x}}$ can be dropped (Exercise 6.1).

Application to the Auxiliary Problem

We now want to use the method of Lagrange multipliers to derive the following lemma.

6.3.2 Lemma. *If $X^*(\mu) \succ 0$ is the optimal solution of the auxiliary problem (6.2), then there is a vector $\tilde{\mathbf{y}} \in \mathbb{R}^m$ such that $X^*(\mu)$ and $\tilde{\mathbf{y}}$ satisfy the equations*

$$A_i \bullet X = b_i, \quad i = 1, 2, \ldots, m,$$

$$C + \mu X^{-1} = \sum_{i=1}^{m} y_i A_i.$$

The proof of this is straightforward, once we have defined suitable functions f and g_i to which we can apply Theorem 6.3.1.

The domain of these functions is the set $R^{n \times n}$ of all $n \times n$ matrices, in which case gradients can conveniently be written as matrices; see the discussion in Sect. 5.4. For the open subset U we use a sufficiently small neighborhood of our optimal solution $X^*(\mu)$. Note that this neighborhood also contains nonsymmetric matrices.

The function f is simply the objective function with the barrier term:

$$f(X) = f_\mu(X) = C \bullet X + \mu \ln \det(X).$$

We have two sets of constraint functions:

$$g_i(X) = A_i \bullet X - b_i, \ i = 1, \ldots, m,$$

for the linear equality constraints of the problem and

$$g_{ij}(X) = x_{ij} - x_{ji}, \ 1 \le i < j \le n,$$

to capture the symmetry of X. Since $X^*(\mu)$ is a maximum of the auxiliary problem (6.2), it is also a (local) maximum subject to only the constraints $g_i(X) = 0$ and $g_{ij}(X) = 0$, because of $X^*(\mu) \succ 0$. Since the g_i and g_{ij} are linear, we need not worry about the regularity of $X^*(\mu)$, and Theorem 6.3.1 can readily be applied. For that, it remains to compute $\nabla f(X)$. The following lemma is the matrix analog of the fact that $(\ln x)' = 1/x$.

6.3.3 Lemma. *For $X \in \mathbb{R}^{n \times n}$ such that $\det X > 0$,*

$$\nabla \ln \det X = (X^T)^{-1}.$$

The proof is left as Exercise 6.2.

Proof of Lemma 6.3.2. The equations $A_i \bullet X^*(\mu) = b_i$, $i = 1, 2, \ldots, m$, follow from the feasibility of $X^*(\mu)$ for the auxiliary problem.

The second set of equations is obtained from the condition involving the Lagrange multipliers. First we need to compute the gradients of the functions f, g_i, g_{ij} defined above. Using Lemma 6.3.3 and the fact that $\nabla(M \bullet X) = M$ for every matrix M, we get

$$\begin{aligned} \nabla f_\mu(X) &= C + \mu(X^T)^{-1}, \\ \nabla g_i(X) &= A_i, \quad i = 1, 2, \ldots, m. \end{aligned}$$

Moreover, $\nabla g_{ij}(X)$ is a skew-symmetric matrix for all $i < j$. (A matrix M is skew-symmetric if $M^T = -M$.) Hence, the method of Lagrange multipliers provides us with $\tilde{y} \in \mathbb{R}^m$ and a skew-symmetric matrix $\tilde{Y} = \sum_{i<j} \tilde{y}_{ij} \nabla g_{ij}(X^*(\mu))$ such that

$$C + \mu(X^*(\mu))^{-1} = \sum_{i=1}^{m} \tilde{y}_i A_i + \tilde{Y},$$

where we have used $X^*(\mu) \in \mathrm{SYM}_n$. Since all matrices in this equation except $\tilde{Y}$ are symmetric, it follows that the skew-symmetric part $\tilde{Y}$ vanishes, and the lemma is proved. $\qquad\square$

We will rewrite the equations developed in the lemma into a more convenient form, by introducing a "slack" matrix $S = \sum_{i=1}^{m} y_i A_i - C = \mu X^{-1}$. Then $X^*(\mu)$ satisfies the *Lagrange system*

$$
\begin{array}{rcll}
A_i \bullet X &=& b_i, & i = 1, 2, \ldots m \\
\sum_{i=1}^{m} y_i A_i - S &=& C \\
SX &=& \mu I_n \\
S, X &\succ& 0
\end{array}
\qquad (6.4)
$$

for suitable $\mathbf{y} \in \mathbb{R}^m$ and $S \in \mathrm{SYM}_n$.

A Primal-Dual Interpretation

From the fact that $X^*(\mu)$ – if it exists – must satisfy the Lagrange system (6.4), we can already deduce that as μ tends to 0, the optimal value $C \bullet X^*(\mu)$ of the barrier problem converges to the value of our original semidefinite program (6.1). This will follow from a stronger property: the equations (6.4) provide a primal feasible solution and a dual feasible solution, with a small *duality gap* (difference between dual and primal objective function value).

6.3.4 Lemma. *If $\tilde{X}, \tilde{S} \in \mathbb{R}^{n \times n}$, $\tilde{y} \in \mathbb{R}^m$ satisfy the Lagrange system (6.4) for some $\mu > 0$, then the following statements hold.*

(i) *The matrix $\tilde{X}$ is a strictly feasible solution of the primal semidefinite program*

$$
\begin{array}{ll}
\text{maximize} & C \bullet X \\
\text{subject to} & A_i \bullet X = b_i, \quad i = 1, 2, \ldots, m \\
& X \succeq 0.
\end{array}
\qquad (6.5)
$$

Here strict feasibility means that $\tilde{X} \succ 0$.

(ii) *The vector $\tilde{y}$ is a strictly feasible solution of the dual semidefinite program*

$$\text{minimize} \quad \mathbf{b}^T \mathbf{y}$$
$$\text{subject to} \quad \sum_{i=1}^{m} y_i A_i - C \succeq 0, \tag{6.6}$$

where strict feasibility means that $\sum_{i=1}^{m} \tilde{y}_i A_i - C \succ 0$.

(iii) The duality gap satisfies

$$\mathbf{b}^T \tilde{\mathbf{y}} - C \bullet \tilde{X} = n\mu.$$

Proof. From $\tilde{S}, \tilde{X} \succ 0$ we immediately obtain that $\tilde{X}$ is strictly feasible for the primal, and $\tilde{\mathbf{y}}$ is strictly feasible for the dual. For the duality gap, we use linearity of $\bullet$ in the first argument to compute

$$
\begin{aligned}
C \bullet \tilde{X} &= \left(\sum_{i=1}^{m} \tilde{y}_i A_i - \tilde{S} \right) \bullet \tilde{X} \\
&= \sum_{i=1}^{m} \tilde{y}_i (A_i \bullet \tilde{X}) - \tilde{S} \bullet \tilde{X} \\
&= \sum_{i=1}^{m} \tilde{y}_i b_i - \tilde{S} \bullet \tilde{X} \\
&= \sum_{i=1}^{m} \tilde{y}_i b_i - \text{Tr}(\underbrace{\tilde{S}\tilde{X}}_{\mu I_n}) \\
&= \mathbf{b}^T \tilde{\mathbf{y}} - n\mu. \qquad \square
\end{aligned}
$$

The lemma just proved shows that if we could compute $X^*(\mu)$ for small μ, then we would have an almost optimal solution of our semidefinite program (6.1). Indeed, since $C \bullet X \leq \mathbf{b}^T \tilde{\mathbf{y}}$ for all feasible solutions X by weak duality (Theorem 4.7.2), $C \bullet X^*(\mu)$ comes to within $n\mu$ of the optimum value.

6.4 Sufficient Conditions for Optimality

So far we have shown that *if* the problem of maximizing $f_\mu(X)$ subject to the constraints $A_i \bullet X = b_i$ and $X \succ 0$ has a maximum at X^*, then there exist $S^* \succ 0$ and $\mathbf{y}^* \in \mathbb{R}^m$ such that $X^*, \mathbf{y}^*, S^*$ satisfy the Lagrange equations (6.4). Next, we formulate conditions on the semidefinite program under which the Lagrange system is uniquely solvable and yields a maximum of f_μ.

For the Lagrange system to be solvable at all, strict feasibility of both the primal and the dual program must be required. The only condition that we will need on top of that is linear independence of the matrices A_i; by Exercise 6.1, this may be assumed without loss of generality.

6.4.1 Lemma. *Suppose that both the primal program (6.5) and the dual program (6.6) have strictly feasible solutions $\tilde{X}$ and $\tilde{\mathbf{y}}$, respectively, and that the matrices A_i, $i = 1, 2, \ldots, m$, are linearly independent (as elements of the vector space SYM_n).*

Then for every $\mu > 0$, the Lagrange system (6.4) has a unique solution $X^* = X^*(\mu)$, $\mathbf{y}^* = \mathbf{y}^*(\mu)$, $S^* = S^*(\mu)$. Moreover, $X^*(\mu)$ is the unique maximizer of f_μ subject to $A_i \bullet X = b_i$, $i = 1, 2, \ldots, m$, and $X \succ 0$.

Proof. Let $\mu > 0$ be fixed. We first show that there is a unique maximizer $X^*(\mu)$ of f_μ. This already implies that the Lagrange system is solved by $X = X^*(\mu)$ with suitable $\mathbf{y}$ and S. Then we prove that there are no other solutions.

We begin with the following claim, which yields existence of $X^*(\mu)$ (uniqueness is then a consequence of strict concavity of f_μ).

Claim. Under the assumptions of the lemma, the set

$$Q = \{X \in \mathrm{SYM}_n : A_i \bullet X = b_i, \ i = 1, 2, \ldots, m, \ X \succ 0, f_\mu(X) \geq f_\mu(\tilde{X})\}$$

is closed and bounded, and hence compact (when interpreted as a subset of $\mathbb{R}^{n^2}$).

Proof of the claim. Closedness is the easy part (Exercise 6.3), the main thing to prove is boundedness. Let $X \succ 0$ satisfy $A_i \bullet X = b_i$, $i = 1, 2, \ldots, m$, and define $\tilde{S} := \sum_{i=1}^{m} \tilde{y}_i A_i - C \succ 0$. As in the proof of Lemma 6.3.4 we compute

$$f_\mu(X) = C \bullet X + \mu \ln \det X = \mathbf{b}^T \tilde{\mathbf{y}} - \tilde{S} \bullet X + \mu \ln \det X.$$

Therefore, $X \in Q$ if and only if

$$\mu \ln \det X - \tilde{S} \bullet X \geq \mu \ln \det \tilde{X} - \tilde{S} \bullet \tilde{X} =: c. \tag{6.7}$$

Next, we want to show that this lower bound implies an *upper* bound on the eigenvalues of every matrix $X \in Q$. Let $\sigma > 0$ be the smallest eigenvalue of $\tilde{S}$, and let $\lambda_1(X), \ldots, \lambda_n(X) > 0$ be the eigenvalues of X. Since $\tilde{S} - \sigma I_n \succeq 0$, we have $(\tilde{S} - \sigma I_n) \bullet X \geq 0$ (a consequence of the self-duality of PSD_n, see Lemma 4.7.5), and hence

$$
\begin{aligned}
\mu \ln \det X - \tilde{S} \bullet X &= \mu \ln \prod_{j=1}^{n} \lambda_j(X) - \tilde{S} \bullet X \\
&\leq \mu \ln \prod_{j=1}^{n} \lambda_j(X) - \sigma \underbrace{I_n \bullet X}_{\mathrm{Tr}(X)} \\
&= \mu \sum_{j=1}^{n} \ln \lambda_j(X) - \sigma \sum_{j=1}^{n} \lambda_j(X),
\end{aligned}
$$

since the trace of a matrix is the sum of its eigenvalues; see Exercise 4.8.

Putting this together with (6.7), we have

$$c \leq \mu \sum_{j=1}^{n} \ln \lambda_j(X) - \sigma \sum_{j=1}^{n} \lambda_j(X), \quad X \in Q. \tag{6.8}$$

The right-hand side is of the form $\sum_{j=1}^{n} h(\lambda_j(X))$, where h is the univariate function $h(x) = \mu \ln x - \sigma x$. Elementary calculus shows that $h(x)$ attains a unique maximum at $x = \mu/\sigma$, and in particular, $h(x)$ is bounded from above. Since (6.8) further yields

$$\sigma \lambda_i(X) - \mu \ln \lambda_i(X) \leq \sum_{j \neq i} h(\lambda_j(X)) - c, \quad i = 1, \ldots, n,$$

we thus know that $\sigma \lambda_i(X) - \mu \ln \lambda_i(X)$ is bounded from above on Q, and since the first term asymptotically dominates the second, $\lambda_i(X)$ itself is bounded from above on Q, for all i. According to Exercise 6.4, Q is then bounded as well. The claim is proved.

We continue with the proof of Lemma 6.4.1. According to the claim, the set Q is nonempty and compact. Hence the continuous function f_μ attains a maximum on it, which, as we know, is unique. This shows that under the assumptions of the lemma, the auxiliary problem has a unique maximum, and by means of Lagrange multipliers we have shown that this maximum yields a solution of the Lagrange system (6.4). It remains to verify that this is the only solution, and only here do we need the linear independence of the A_i.

To this end, we show that for every solution $\tilde{X}', \tilde{y}', \tilde{S}'$ of (6.4), $\tilde{X}'$ also maximizes f_μ, hence $\tilde{X}' = X^*(\mu)$. We note that $\tilde{S}'$ and $\tilde{y}'$ are uniquely determined by $\tilde{X}'$ through the relations $SX = \mu I_n$ and $\sum_{i=1}^{m} y_i A_i - S = C$ from (6.4), using the assumption that the A_i are linearly independent. Thus, unique solvability of the Lagrange system follows.

To show that $\tilde{X}'$ maximizes f_μ in the auxiliary problem, we use a fact established in Exercise 6.5: the method of Lagrange multipliers as outlined on page 103 sometimes also provides *sufficient* conditions for a local maximum at $\tilde{x}$ – namely, in the case where the function f is concave in a small neighborhood of $\tilde{x}$ and the g_i are linear. To apply this, we recall that if $\tilde{X}', \tilde{y}', \tilde{S}'$ solve the Lagrange system (6.4), then $\tilde{X}'$ is a feasible solution of the problem

$$\begin{aligned}\text{Maximize} \quad & f_\mu(X) := C \bullet X + \mu \ln \det X \\ \text{subject to} \quad & A_i \bullet X = b_i, \quad i = 1, 2, \ldots m \\ & X \in \text{SYM}_n,\end{aligned}$$

and $\tilde{y}'_1, \tilde{y}'_2, \ldots, \tilde{y}'_m$ are Lagrange multipliers for $\tilde{X}'$. This is how we derived the Lagrange system.

Since $\tilde{X}' \succ 0$, the function f_μ is concave in a small neighborhood of $\tilde{X}'$ (Lemma 6.2.2), and hence $\tilde{X}'$ is a local maximum of f_μ on the set of positive definite matrices, by sufficiency of Lagrange multipliers mentioned above.

The strict concavity of f_μ also implies that every local maximum coincides with the unique global maximum (the argument is similar to the one that we used to establish uniqueness of the global maximum). Hence $\tilde{X}'$ indeed maximizes f_μ in the auxiliary problem, and the lemma is proved. $\qquad\square$

6.5 Following the Central Path

In this section we address the question of how the Lagrange system (6.4) can be solved for small μ, since this is what we need to get good primal and dual solutions; see Lemma 6.3.4.

Let us define the *primal-dual central path* of the semidefinite program (6.5) as the set

$$\left\{ (X^*(\mu), \mathbf{y}^*(\mu), S^*(\mu)) \in \mathrm{PSD}_n \times \mathbb{R}^m \times \mathrm{PSD}_n : \mu > 0 \right\}.$$

We can indeed call this a path, since $X^*(\mu), \mathbf{y}^*(\mu)$ and $S^*(\mu)$ are continuous functions of μ (Exercise 6.6).

The idea of the central path method is to start at some $(\tilde{X}, \tilde{y}, \tilde{S})$ lying close to the central path, and approximately follow the central path until μ becomes sufficiently small.

Let us fix μ for now and introduce a *central path function* F capturing the deviation of a given triple $(X, \mathbf{y}, S)$ from the central path:

$$F \colon \mathrm{SYM}_n \times \mathbb{R}^m \times \mathrm{SYM}_n \to \mathbb{R}^m \times \mathrm{SYM}_n \times \mathrm{SYM}_n,$$

$$F(X, \mathbf{y}, S) = \begin{pmatrix} P(X, \mathbf{y}, S) \\ Q(X, \mathbf{y}, S) \\ R(X, \mathbf{y}, S) \end{pmatrix}, \tag{6.9}$$

with

$$P(X, \mathbf{y}, S) \;=\; \begin{pmatrix} A_1 \bullet X - b_1 \\ A_2 \bullet X - b_2 \\ \vdots \\ A_m \bullet X - b_m \end{pmatrix},$$

$$Q(X, \mathbf{y}, S) \;=\; \sum_{i=1}^m y_i A_i - S - C,$$

$$R(X, \mathbf{y}, S) \;=\; SX - \mu I_n.$$

We know that $F(X^*(\mu), \mathbf{y}^*(\mu), S^*(\mu)) = (\mathbf{0}, \mathbf{0}, 0)$, and that this is the only zero of F subject to $X, S \succ 0$, by the unique solvability of the Lagrange system. Furthermore, we would like to compute this zero for small μ, in order to obtain almost optimal solutions for the primal and dual programs (6.5) and (6.6), as guaranteed by Lemma 6.3.4.

Directly solving the system $F(X, \mathbf{y}, S) = \mathbf{0}$ is difficult, since it contains the n^2 nonlinear equations $SX - \mu I_n = 0$. But there is a well-known stepwise method for computing the zero of a function that is based on solving systems of *linear* equations only. This method is called *Newton's method*.

Newton's Method

Let us first recall Newton's method for finding a zero of a differentiable univariate function f. We start with some initial guess $x^{(0)}$, and given $x^{(i)}$, $i \geq 0$, we set

$$x^{(i+1)} := x^{(i)} - \frac{f(x^{(i)})}{f'(x^{(i)})},$$

or equivalently,

$$f'(x^{(i)}) \left(x^{(i+1)} - x^{(i)} \right) = -f(x^{(i)}). \tag{6.10}$$

Under suitable conditions on f and the initial guess, the sequence $(x^{(i)})_{i \in \mathbb{N}}$ quickly converges to a zero of f.

Newton's method generalizes to higher dimensions. If $f : \mathbb{R}^n \to \mathbb{R}^n$, then (6.10) becomes

$$Df(\mathbf{x}^{(i)}) \left(\mathbf{x}^{(i+1)} - \mathbf{x}^{(i)} \right) = -f(\mathbf{x}^{(i)}), \tag{6.11}$$

where $Df(\mathbf{x})$ is the *Jacobian* of f at $\mathbf{x}$, that is, the $n \times n$ matrix given by

$$(Df(\mathbf{x}))_{ij} = \frac{\partial f(\mathbf{x})_i}{\partial x_j}.$$

For this to work, the matrix $Df(\mathbf{x}^{(i)})$ must be invertible, for otherwise, the next iterate $\mathbf{x}^{(i+1)}$ is not well-defined.

Newton's Method Applied to the Central Path Function

Let us now derive the formulas for one step of Newton's method applied to the central path function F. For this, we ignore the symmetry of X and S and write $F : \mathbb{R}^{2n^2 + m} \to \mathbb{R}^{2n^2 + m}$. The calculations itself are somewhat boring, but since they involve one or two generally useful arguments, we will do them anyway.

As in Lemma 6.4.1, we assume that we have $X^{(i)} = \tilde{X}$, $\mathbf{y}^{(i)} = \tilde{\mathbf{y}}$, $S^{(i)} = \tilde{S}$ such that

$$A_i \bullet \tilde{X} = b_i, \ i = 1, 2, \ldots, m, \quad \sum_{i=1}^{m} \tilde{y}_i A_i - \tilde{S} = C, \quad \tilde{S}, \tilde{X} \succ 0.$$

Now we want to compute the next iterate $X^{(i+1)} = \tilde{X}'$, $\mathbf{y}^{(i+1)} = \tilde{\mathbf{y}}'$, $S^{(i+1)} = \tilde{S}'$ according to the general recipe in (6.11). Let us write

$$\Delta X = \tilde{X}' - \tilde{X}, \quad \Delta \mathbf{y} = \tilde{\mathbf{y}}' - \tilde{\mathbf{y}}, \quad \Delta S = \tilde{S}' - \tilde{S}. \tag{6.12}$$

Then (6.11) becomes

$$DF(\tilde{X}, \tilde{\mathbf{y}}, \tilde{S}) \begin{pmatrix} \Delta X \\ \Delta \mathbf{y} \\ \Delta S \end{pmatrix} = -F(\tilde{X}, \tilde{\mathbf{y}}, \tilde{S}) = \begin{pmatrix} \mathbf{0} \\ 0 \\ \mu I_n - \tilde{S}\tilde{X} \end{pmatrix}. \tag{6.13}$$

The matrix $DF(X, \mathbf{y}, S)$ has the following block structure.

$$DF(X, \mathbf{y}, S) = \left(\begin{array}{c|c|c} DP_{\mathbf{y},S}(X) & 0 & 0 \\ \hline 0 & DQ_{X,S}(\mathbf{y}) & DQ_{X,\mathbf{y}}(S) \\ \hline DR_{\mathbf{y},S}(X) & 0 & DR_{X,\mathbf{y}}(S) \end{array} \right),$$

where the subscripts mean that the corresponding arguments are fixed. After dropping constant terms (which does not change derivatives), all the five "single-argument" functions that we need to differentiate in the blocks are linear. Exercise 6.7 shows that for a linear function F, we have $DF(\mathbf{x})\mathbf{y} = F(\mathbf{y})$, and we can use this formula to compute the left-hand side of (6.13) componentwise:

$$DP_{\mathbf{y},S}(X)(\Delta X) = \begin{pmatrix} A_1 \bullet (\Delta X) \\ A_2 \bullet (\Delta X) \\ \vdots \\ A_m \bullet (\Delta X) \end{pmatrix},$$

$$DQ_{X,S}(\mathbf{y})(\Delta \mathbf{y}) + DQ_{X,\mathbf{y}}(S)(\Delta S) = \sum_{i=1}^{m} (\Delta \mathbf{y})_i A_i - \Delta S$$

$$DR_{\mathbf{y},S}(X)(\Delta X) + DR_{X,\mathbf{y}}(S)(\Delta S) = S(\Delta X) + (\Delta S)X.$$

Hence, (6.13) is the following system of *linear* equations for $\Delta X, \Delta \mathbf{y}, \Delta S$.

$$A_i \bullet (\Delta X) = \mathbf{0}, \quad i = 1, 2, \ldots, m \qquad (6.14)$$

$$\sum_{i=1}^{m} (\Delta \mathbf{y})_i A_i - (\Delta S) = 0 \qquad (6.15)$$

$$\tilde{S}(\Delta X) + (\Delta S)\tilde{X} = \mu I_n - \tilde{S}\tilde{X} \qquad (6.16)$$

We claim that this system has a unique solution $(\Delta X, \Delta \mathbf{y}, \Delta S)$. To see this, we start by solving the last equation (6.16) for ΔX:

$$\begin{aligned}
\Delta X &= \tilde{S}^{-1}(\mu I_n - \tilde{S}\tilde{X} - (\Delta S)\tilde{X}) \\
&= \tilde{S}^{-1}(\mu I_n - \tilde{S}\tilde{X} - \sum_{i=1}^{m} (\Delta \mathbf{y})_i A_i \tilde{X}) \\
&= \mu \tilde{S}^{-1} - \tilde{X} - \tilde{S}^{-1} \sum_{i=1}^{m} (\Delta \mathbf{y})_i A_i \tilde{X}, \qquad (6.17)
\end{aligned}$$

using the second equation (6.15). Substituting this into the first equation set (6.14) yields a system only for $\Delta \mathbf{y}$:

$$A_i \bullet (\mu \tilde{S}^{-1} - \tilde{X} - \tilde{S}^{-1} \sum_{i=1}^{m} (\Delta \mathbf{y})_i A_i \tilde{X}) = \mathbf{0}, \quad i = 1, 2, \ldots, m.$$

Equivalently,

$$A_i \bullet (\tilde{S}^{-1} \sum_{i=1}^{m} (\Delta \mathbf{y})_i A_i \tilde{X}) = \mu A_i \bullet \tilde{S}^{-1} - b_i, \; i = 1, \ldots, m.$$

Exercise 6.8 asks you to prove that this is a system of the form

$$M(\Delta \mathbf{y}) = \mathbf{q}, \quad M \succ 0;$$

in particular, M is invertible. It follows that $\Delta \mathbf{y}$ is uniquely determined, and substituting back into (6.15) yields ΔS, from which we in turn obtain ΔX through (6.17).

Making ΔX Symmetric

We recall that we would like to obtain the next iterate in Newton's method (which hopefully brings us closer to the zero of $F(X, \mathbf{y}, S)$) via $\tilde{X}' = \tilde{X} + \Delta X$, $\tilde{\mathbf{y}}' = \tilde{\mathbf{y}} + \Delta \mathbf{y}$, $\tilde{S}' = \tilde{S} + \Delta S$. The only problem is that ΔX may not be symmetric, in which case $(\tilde{X}', \tilde{\mathbf{y}}', \tilde{S}')$ is not a valid next iterate. We note that ΔS is symmetric as a consequence of (6.15).

The problem with ΔX is due to the fact that we have ignored the symmetry constraints and solved the system (6.13) for $X, S \in \mathbb{R}^{n \times n}$ instead of $X, S \in \mathrm{SYM}_n$.

A simple fix is to update according to a "symmetrized" ΔX:

$$\tilde{X}' = \tilde{X} + \frac{1}{2} \left(\Delta X + (\Delta X)^T \right).$$

It can be shown that this *modified Newton step* also leads to theoretical convergence and good practical performance [HRVW96].

But to get polynomial runtime bounds, we need to proceed differently. It is possible to define a function $F'(X, \mathbf{y}, S)$, actually in various ways, for which the Newton step yields a symmetric update matrix ΔX. This is discussed in detail in [MT00, Sect. 10.3].

One possible choice is the following. Let

$$S^{(\tilde{X})}(M) := \frac{1}{2} \left(\tilde{X}^{-1/2} M \tilde{X}^{1/2} + (\tilde{X}^{-1/2} M \tilde{X}^{1/2})^T \right),$$

where $\tilde{X}^{1/2}$ is the square root of $\tilde{X}$, the unique positive definite matrix whose square is $\tilde{X}$ (the existence of $\tilde{X}^{1/2}$ follows from diagonalization: if $\tilde{X} = U^T D U$ with U orthogonal and D diagonal, we can set $\tilde{X}^{1/2} := U^T \sqrt{D} U$, where $\sqrt{D}$ is the diagonal matrix obtained from D by taking the square roots of all diagonal elements).

Instead of $R(X, \mathbf{y}, S) = SX - \mu I_n$, we now use the function

$$R^{(\tilde{X})}(X, \mathbf{y}, S) = S^{(\tilde{X})}(XS) - \mu I_n$$

in the definition (6.9) of the function F. Reintroducing μ, let us call the resulting function $F_\mu^{(\tilde{X})}$.

When we compute the Newton system (6.13) for $F_\mu^{(\tilde{X})}$, we again arrive at (6.14) and (6.15), but (6.16) gets replaced with

$$\tilde{X}^{-1/2}(\tilde{X}\Delta S + \Delta X \tilde{S})\tilde{X}^{1/2} + \tilde{X}^{1/2}(\Delta S \tilde{X} + \tilde{S}\Delta X)\tilde{X}^{-1/2}$$
$$= 2(\mu I_n - \tilde{X}^{1/2}\tilde{S}\tilde{X}^{1/2}). \quad (6.18)$$

This yields a symmetric ΔX [Mon97]; the matrix square roots look somewhat scary, but they can be avoided: the modified Newton system is equivalent to a "square-root-free" system of linear equations that was derived by Kojima et al. [KSH97], see [Mon97, Lemma 2.1].

But there is a price to pay for a symmetric ΔX: We had to change the function F, losing the crucial property that a zero $(X^*, \mathbf{y}^*, S^*)$ of F satisfies $S^* X^* = \mu I_n$. Actually, our modified function $F_\mu^{(\tilde{X})}$ even depends on the current iterate $\tilde{X}$, so the next Newton step will be with respect to a different function. But this is not a problem: If for *any* $\tilde{X}$, the triple $(X^*, \mathbf{y}^*, S^*)$ is a zero of $F_\mu^{(\tilde{X})}$, we still have

$$S^* \bullet X^* = \mu n$$

(Exercise 6.9). If we reexamine the proof of Lemma 6.3.4, we see that this is all we need in order to get small duality gap.

The Algorithm

Given that we know how to perform one step of Newton's method, the following seems to be a natural way of simultaneously solving the semidefinite program (6.5) and its dual (6.6) up to duality gap $\varepsilon > 0$. Choose $\mu = \varepsilon/n$ and then perform Newton steps on $F_\mu^{\tilde{X}}$ ($\tilde{X}$ always the current iterate), until $F_\mu^{\tilde{X}} \approx \mathbf{0}$. Then the current $\tilde{X}, \tilde{\mathbf{y}}$ are almost optimal solutions of (6.5) and (6.6) with duality gap ε.

There are three obstacles to overcome. First, recall that we are attempting to find the unique solution of the Lagrange system (6.4), and this includes the constraints $X, S \succ 0$. But Newton's method does not know about them and cannot guarantee $X^{(i)}, S^{(i)} \succ 0$ even if this holds for $i = 0$. The second problem is that a *fast* convergence can be proved only if we start sufficiently close to the central path. Our initial solution $(\tilde{X}, \tilde{\mathbf{y}}, \tilde{S})$ could be too far away. The third and very mundane problem is that we may not even have a feasible solution to begin with.

The actual algorithm addresses all these obstacles and consists of the following two phases, of which we only describe the second one in detail.

Self-dual Embedding. In order for any interior-point method to get started, we need – well – some interior point. In the case of the primal-dual central path algorithm, an interior point can be obtained from strictly feasible solutions $\tilde{X}$ for the primal problem (6.5) and $\tilde{\mathbf{y}}$ for the dual problem (6.6). The idea is to embed both the primal problem and the dual problem into a "larger" (and self-dual) semidefinite program for which an interior point is readily available. Solving this larger problem using the primal-dual central path algorithm then also yields (approximately) optimal primal and dual solutions for the original problem; see [WSV00, Chap. 5] for details.

Path following. Suppose that our interior point is close enough to the central path for some (not necessarily small) value of μ. For the *long-step path following method* [Mon97, Sect. 5], any interior point is close enough, while the faster *short-step path following method* [Mon97, Sect. 4] that we detail below requires a concrete distance bound. In both cases, it would not be good to perform Newton steps with the current (possibly large) value of μ fixed. All we would get is a point arbitrarily close to the central path at μ, but unless μ is small, this is not an approximately optimal solution to the semidefinite program.

Instead, the approach is as follows. In each iteration of Newton's method, we perform the step with respect to a *slightly smaller* value of μ. The intention is to bring the iterate closer to the central path at the smaller μ, and then to repeat the process until the current iterate is close to the central path for sufficiently small μ. Here is our notion of being close.

6.5.1 Definition. *For a real number $\gamma > 0$, the γ-neighborhood of the central path is the set of interior points $(X, \mathbf{y}, S)$ such that $\|X^{1/2}SX^{1/2} - \mu I_n\|_F \leq \mu\gamma$.*

We recall that $\|\cdot\|_F$ is the Frobenius norm of a matrix. It is justified to call this a neighborhood of the central path: indeed, a point on the central path satisfies $SX = \mu I_n$, and this implies $X^{1/2}SX^{1/2} = \mu I_n$, so the point is in all neighborhoods.

Here is a generic step of the short-step path following algorithm:

---

1. Given the current iterate $X^{(i)}, \mathbf{y}^{(i)}, S^{(i)}$, set

$$\mu_i := \frac{S^{(i)} \bullet X^{(i)}}{n}.$$

   (If $(X^{(i)}, \mathbf{y}^{(i)}, S^{(i)})$ is on the central path, then $X^{(i)} = X^*(\mu_i)$.)

2. Perform one step of Newton's method w.r.t. $F_\mu^{X^{(i)}}$, where

$$\mu := \sigma\mu_i, \text{ and } \sigma := 1 - \frac{0.3}{\sqrt{n}}$$

   is the *centrality parameter*. This means, compute $\Delta X, \Delta\mathbf{y}, \Delta S$ by solving (6.14), (6.15) and (6.18), and set

$$
\begin{aligned}
X^{(i+1)} &:= X^{(i)} + \Delta X, \\
\mathbf{y}^{(i+1)} &:= \mathbf{y}^{(i)} + \Delta\mathbf{y}, \\
S^{(i+1)} &:= S^{(i)} + \Delta S.
\end{aligned}
$$

6.5.2 Theorem ([Mon97]). *Let $\gamma := 0.3$, and suppose that $(X^{(i)}, \mathbf{y}^{(i)}, S^{(i)})$ is an interior point in the γ-neighborhood of the central path. Then*

$$(X^{(i+1)}, \mathbf{y}^{(i+1)}, S^{(i+1)})$$

is again an interior point in the γ-neighborhood of the central path, and

$$S^{(i+1)} \bullet X^{(i+1)} = \sigma \cdot S^{(i)} \bullet X^{(i)}.$$

A bound on the runtime directly follows from the geometric decrease of the sequence $(S^{(i)} \bullet X^{(i)})_{i \in \mathbb{N}}$.

6.5.3 Corollary. *Let $\varepsilon > 0$ be given, and let us suppose that we have $(\tilde{X}, \tilde{\mathbf{y}}, \tilde{S})$ such that*

$$A_i \bullet \tilde{X} = b_i, \ i = 1, 2, \ldots, m, \ \tilde{X} \succ 0 \quad \text{(strict primal feasibility)},$$

$$\sum_{i=1}^{m} \tilde{y}_i A_i - \tilde{S} = C, \ \tilde{S} \succ 0 \quad \text{(strict dual feasibility)},$$

and $(\tilde{X}, \tilde{y}, \tilde{S})$ is in the 0.3-neighborhood of the central path. Let $\sigma = 1 - 0.3/\sqrt{n}$ and

$$k \geq \log_{1/\sigma} \frac{\tilde{S} \bullet \tilde{X}}{\varepsilon} = O\left(\sqrt{n} \log \frac{\tilde{S} \bullet \tilde{X}}{\varepsilon}\right).$$

Then the k-th iterate in the above algorithm started with $X^{(0)} = \tilde{X}, \mathbf{y}^{(0)} = \tilde{\mathbf{y}}$, $S^{(0)} = \tilde{S}$ satisfies

(i) $A_i \bullet X^{(k)} = b_i, \ i = 1, 2, \ldots, m, \ X^{(k)} \succ 0$
(ii) $\sum_{i=1}^{m} \mathbf{y}_i^{(k)} - S^{(k)} = C, \ S^{(k)} \succ 0$
(iii) $S^{(k)} \bullet X^{(k)} \leq \varepsilon$

We recall that the condition $S^{(k)} \bullet X^{(k)} \leq \varepsilon$ leads to small duality gap, and therefore to a primal solution that is optimal up to an additive error of ε (see the proof of Lemma 6.3.4).

We stress that on top of requiring strictly feasible primal and dual points, we also need to require that the initial solution is close to the central path. As already indicated, this assumption can be removed by using the long-step method, but then an extra $O(n)$ factor enters the runtime bound [Mon97, Corollary 5.3].

Also, good (polynomial) runtime bounds result only if $S^{(0)} \bullet X^{(0)}$ is not too large. This is the case in many practical applications, but in the worst case it cannot be guaranteed; see, e.g., the pathological semidefinite program at the end of Sect. 2.6.

Conclusion. The main virtue of interior-point methods for semidefinite programming is that they are easy to implement and work well in practice.

Generally they are not easy to analyze. Specific algorithms with known worst-case bounds are often rather slow in practice, and they are replaced by other variants, which are apparently faster but whose theoretical performance is unknown. For example, the algorithm of Helmberg [HRVW96] has been shown to converge, but without any bounds on the convergence rate. Still, Helmberg is offering an efficient semidefinite programming solver based on his algorithm; the convergence is fast in practice, and the primitive operations are very simple.

From a theoretical point of view, this situation may be somewhat unsatisfactory. But the main message of this chapter is this: *Semidefinite programs can efficiently be solved in theory and in practice, using interior-point methods,* and we have outlined how this works.

Exercises

6.1 Consider the problem of maximizing a function $f(\mathbf{x})$ (with continuous partial derivatives) subject to linear constraints $g_i(\mathbf{x}) = \mathbf{a}_i\mathbf{x} - b_i = 0$, $i = 1, \ldots, m$, (here, the $\mathbf{a}_i$ are row vectors). Show that if $\tilde{\mathbf{x}}$ is a maximizer of f subject to the constraints $g_i(\mathbf{x}) = 0$, then there exists $\mathbf{y} \in \mathbb{R}^m$ such that

$$\nabla f(\tilde{\mathbf{x}}) = \sum_{i=1}^{m} y_i \mathbf{a}_i.$$

In particular, we do not need the requirement that $\tilde{\mathbf{x}}$ is a regular point.

Hint: You may assume correctness of the general method of Lagrange multipliers as outlined on page 103.

6.2 Prove that for $X \in \mathbb{R}^{n \times n}$, $\nabla \ln \det(X) = (X^T)^{-1}$. Here the gradient is the matrix $G = (g_{ij})$ of partial derivatives:

$$g_{ij} = \frac{\partial \nabla \ln \det(X)}{\partial x_{ij}}, \quad 1 \leq i, j \leq n.$$

6.3 With f_μ as defined in (6.2), prove that for every real number r, the set

$$\{X \in \mathrm{SYM}_n : A_i \bullet X = b_i, i = 1, 2, \ldots, m, \ X \succ 0, f_\mu(X) \geq r\}$$

is closed.

6.4 Let $Q \subseteq \mathrm{PSD}_n$ be a set of matrices for which all eigenvalues are bounded by some global constant c. Prove that Q is bounded as well (in Frobenius norm).

6.5 Prove that if $\mathbf{x}$ and $y_1, \ldots, y_m$ satisfy (6.3), where the g_i are linear functions $g_i(\mathbf{x}) = \mathbf{a}_i\mathbf{x} - b_i$, and f is concave in a small neighborhood U of $\mathbf{x}$, then $\mathbf{x}$ is a local maximum of f subject to $g_i(\mathbf{x}) = 0$, $i = 1, \ldots, m$.

6.6 Prove that the function

$$\mu \mapsto (X^*(\mu), \mathbf{y}^*(\mu), S^*(\mu))$$

that maps μ to the unique solution of (6.4) is continuous.

6.7 Let $F: \mathbb{R}^k \to \mathbb{R}^k$ be a linear function, and let $\mathbf{x}, \mathbf{y} \in \mathbb{R}^k$. Prove that

$$DF(\mathbf{x})\mathbf{y} = F(\mathbf{y}).$$

6.8 Let $\tilde{S}, \tilde{X} \succ 0$, and let A_i be linearly independent matrices. Prove that there is a matrix $M \succ 0$ such that

$$A_i \bullet (\tilde{S}^{-1} \sum_{i=1}^{m} y_i A_i \tilde{X}) = M\mathbf{y} \text{ for all } \mathbf{y} \in \mathbb{R}^m.$$

6.9 Let P be a fixed invertible matrix and consider the function

$$S(M) = \frac{1}{2} \left(PMP^{-1} + (PMP^{-1})^T \right).$$

Prove that $S(M) = Q$ implies $\text{Tr}(M) = \text{Tr}(Q)$.

Chapter 7
Copositive Programming

In this chapter we come back to the topic of cone programming (Chap. 4). So far, we have seen two important classes of cone programs, namely linear and semidefinite programs. Both classes are "easy" in the sense that there are practically efficient algorithms and polynomial-time complexity results, at least under certain conditions (in the semidefinite case).

We cannot expect similar results for general cone programs, since the involved closed convex cone(s) may be "hard." In this chapter we exhibit two concrete hard cones in SYM_n: the cone of *completely positive* matrices, and its dual, the cone of *copositive* matrices. The semidefinite cone PSD_n is wedged between them.

On the way we will encounter some connections to the material on the theta function in Chap. 3, and cone programming duality will appear in an interesting (not self-dual) way.

As it turns out, copositive matrices naturally arise in the context of (locally) minimizing smooth functions, and a hardness results for the latter problem will easily follow from the hardness of copositive programming.

The material in this chapter is classic; our presentation mostly follows Laurent and Rendl [LR05] as well as Murty and Kabadi [MK87].

7.1 The Copositive Cone and Its Dual

Let us start with a matrix class that is closely related to the class of positive semidefinite matrices.

7.1.1 Definition. *A matrix $M \in \mathrm{SYM}_n$ is called copositive if*

$$\mathbf{x}^T M \mathbf{x} \geq 0 \text{ for all } \mathbf{x} \geq 0.$$

Every positive semidefinite matrix is also copositive due to Fact 2.2.1(ii), but the converse is false. For example, every matrix with only nonnegative

B. Gärtner and J. Matoušek, *Approximation Algorithms and Semidefinite Programming*, DOI 10.1007/978-3-642-22015-9_7,
© Springer-Verlag Berlin Heidelberg 2012

entries is copositive. Hence

$$M = \begin{pmatrix} 0 & 1 \\ 1 & 0 \end{pmatrix}$$

is copositive but not positive semidefinite because $\det(M) = -1$.

7.1.2 Definition. *Let*

$$\mathrm{COP}_n := \{M \in \mathrm{SYM}_n : \mathbf{x}^T M \mathbf{x} \geq 0 \text{ for all } \mathbf{x} \geq 0\}$$

be the set of copositive matrices in SYM_n.

The proof of the following lemma is exactly the same as the proof of Lemma 4.2.2 for the positive semidefinite case, and so we omit it.

7.1.3 Lemma. *The set* COP_n *is a closed convex cone in* SYM_n.

Once we have a closed convex cone, it is a natural reflex to compute its dual cone. We recall that for a cone $K \subseteq \mathrm{SYM}_n$, the dual cone is

$$K^* = \{Y \in \mathrm{SYM}_n : Y \bullet X \geq 0 \text{ for all } X \in K\}.$$

From the equation

$$\mathbf{x}^T M \mathbf{x} = M \bullet \mathbf{x}\mathbf{x}^T, \tag{7.1}$$

which we used before (in Sect. 4.8), it follows that all matrices of the form $\mathbf{x}\mathbf{x}^T$ with $\mathbf{x} \geq 0$ (and finite sums of such matrices) are in COP_n^*. Let us give a name to such matrices.

7.1.4 Definition. *A matrix* $M \in \mathrm{SYM}_n$ *is called* completely positive *if for some* ℓ, *there are* ℓ *nonnegative vectors* $\mathbf{x}_1, \mathbf{x}_2, \ldots, \mathbf{x}_\ell \in \mathbb{R}_+^n$, *such that*

$$M = \sum_{i=1}^{\ell} \mathbf{x}_i \mathbf{x}_i^T = AA^T, \tag{7.2}$$

where $A \in \mathbb{R}^{n \times \ell}$ *is the (nonnegative) matrix with columns* $\mathbf{x}_1, \mathbf{x}_2, \ldots, \mathbf{x}_\ell$.

Every completely positive matrix is positive semidefinite. Indeed, M is positive semidefinite if and only if M is of the form (7.2) for $\mathbf{x}_1, \mathbf{x}_2, \ldots,$ $\mathbf{x}_\ell \in \mathbb{R}^n$; see Lemma 4.7.4. In the definition, ℓ is arbitrary, but we do not need to consider $\ell > \binom{n+1}{2}$. The proof of the following lemma is left as Exercise 7.1.

7.1.5 Lemma. M *is completely positive if and only if there are* $\binom{n+1}{2}$ *nonnegative vectors* $\mathbf{x}_1, \mathbf{x}_2, \ldots, \mathbf{x}_{\binom{n+1}{2}} \in \mathbb{R}^n$ *such that*

$$M = \sum_{i=1}^{\binom{n+1}{2}} \mathbf{x}_i \mathbf{x}_i^T.$$

We already know that COP_n^* contains the set of completely positive matrices, but in fact, equality holds. To prove this, let us first show that the completely positive matrices form a closed convex cone as well.

7.1.6 Lemma. *The set*

$$\mathrm{POS}_n := \{M \in \mathrm{SYM}_n : M \text{ is completely positive}\}$$

is a closed convex cone, and we have $\mathrm{POS}_n \subseteq \mathrm{PSD}_n \subseteq \mathrm{COP}_n$.

Proof. The latter chain of inclusions writes out two observations that we have made before. To show that POS_n is a cone, we first observe that if $M = \sum_{i=1}^{\ell} \mathbf{x}_i \mathbf{x}_i^T \in \mathrm{POS}_n$, then also $\lambda M = \sum_{i=1}^{\ell} (\sqrt{\lambda}\mathbf{x}_i)(\sqrt{\lambda}\mathbf{x}_i)^T \in \mathrm{POS}_n$ for $\lambda \geq 0$. For $M, M' \in \mathrm{POS}_n$, $M + M' \in \mathrm{POS}_n$ is immediate from Definition 7.1.4. It remains to prove that the cone POS_n is closed. Here we need that ℓ can be bounded; see Lemma 7.1.5. Let $(M^{(k)})_{k \in \mathbb{N}}$ be a sequence such that

$$M^{(k)} = \sum_{i=1}^{\binom{n+1}{2}} \mathbf{x}_i^{(k)} \mathbf{x}_i^{(k)\,T} = A^{(k)} A^{(k)\,T} \in \mathrm{POS}_n$$

for all k and $\lim_{k \to \infty} M^{(k)} = M \in \mathrm{SYM}_n$. We need to show that $M \in \mathrm{POS}_n$.

Let the vector $\mathbf{a}_i^{(k)} \in \mathbb{R}_+^{\binom{n+1}{2}}$ denote the i-th column of $A^{(k)\,T}$, $i = 1, \ldots, n$. Thus,

$$m_{ii} = \lim_{k \to \infty} M_{ii}^{(k)} = \lim_{k \to \infty} \mathbf{a}_i^{(k)\,T} \mathbf{a}_i^{(k)} = \lim_{k \to \infty} \|\mathbf{a}_i^{(k)}\|^2, \quad i = 1, 2, \ldots, n,$$

which implies that the sequence of vectors $(\mathbf{a}_i^{(k)})_{k \in \mathbb{N}}$ is bounded. Hence there is a convergent subsequence with limit $\mathbf{a}_i$. This yields $\mathbf{a}_i \geq \mathbf{0}$ and $m_{ii} = \|\mathbf{a}_i\|^2$, by continuity of $\mathbf{a}_i^{(k)} \mapsto \|\mathbf{a}_i^{(k)}\|^2$, and since any subsequence of the convergent sequence $(\|\mathbf{a}_i^{(k)}\|^2)_{k \in \mathbb{N}}$ has the same limit. Furthermore,

$$m_{ij} = \lim_{k \to \infty} \mathbf{a}_i^{(k)\,T} \mathbf{a}_j^{(k)} = \mathbf{a}_i^T \mathbf{a}_j$$

by the same kind of argument, so that

$$M = AA^T,$$

where A^T is the nonnegative matrix with columns $\mathbf{a}_1, \mathbf{a}_2, \ldots, \mathbf{a}_n$. So M is completely positive indeed. $\qquad\square$

Now we can prove duality between COP_n and POS_n.

7.1.7 Theorem. $\mathrm{POS}_n^* = \mathrm{COP}_n$.

Proof. To prove $\mathrm{COP}_n \subseteq \mathrm{POS}_n^*$, we fix $M \in \mathrm{COP}_n$ and show that $M \bullet X \geq 0$ for all $X \in \mathrm{POS}_n$. To this end we calculate

$$\underbrace{M}_{\in \mathrm{COP}_n} \bullet \underbrace{\sum_{i=1}^{\ell} \mathbf{x}_i \mathbf{x}_i^T}_{\in \mathrm{POS}_n} = \sum_{i=1}^{\ell} M \bullet \mathbf{x}_i \mathbf{x}_i^T \overset{(7.1)}{=} \sum_{i=1}^{\ell} \mathbf{x}_i^T M \underbrace{\mathbf{x}_i}_{\geq 0} \geq 0.$$

For $\mathrm{COP}_n \supseteq \mathrm{POS}_n^*$, consider $M \notin \mathrm{COP}_n$. Then there is $\tilde{\mathbf{x}} \geq 0$ such that $\tilde{\mathbf{x}}^T M \tilde{\mathbf{x}} < 0$. Using (7.1) again, $\tilde{\mathbf{x}}\tilde{\mathbf{x}}^T \in \mathrm{POS}_n$ witnesses $M \notin \mathrm{POS}_n^*$. $\square$

In order to get some intuition on COP_n and POS_n, Exercise 7.2 asks you to characterize these cones for $n = 2$.

7.2 A Copositive Program for the Independence Number of a Graph

In this section we want to show that there is a copositive program whose value is the independence number of a given graph, the size of a maximum independent set. A *copositive program* looks just like a semidefinite program, except that it has the constraint $X \in \mathrm{COP}_n$ or $X \in \mathrm{POS}_n$ rather than $X \succeq 0$. The result implies that copositive programming is NP-hard in general, and we will draw some further consequences from this in Sect. 7.3 below.

Throughout this section, we fix a graph $G = (V, E)$ with $V = \{1, 2, \ldots, n\}$, $n \geq 1$. We also recall that $\overline{E}$ denotes the edges of the complementary graph $\overline{G}$.

7.2.1 Theorem. *The copositive program*

$$\begin{array}{ll} \textit{minimize} & t \\ \textit{subject to} & y_{ij} = -1, \ \textit{if } \{i,j\} \in \overline{E} \\ & y_{ii} = t - 1, \ \textit{for all } i = 1, 2, \ldots, n \\ & Y \in \mathrm{COP}_n \end{array} \qquad (7.3)$$

has value $\alpha(G)$, the size of a maximum independent set in G.

Before we set out to prove this, let us discuss the relation to Chap. 3. The attentive reader may remember the very similar program (3.7) whose value is the theta function $\vartheta(G)$. Program (7.3) is simply the relaxation of (3.7), obtained by replacing the constraint $Y \succeq 0$ with $Y \in \mathrm{COP}_n$.

Thus we know that the value of (7.3) is at most $\vartheta(G)$. Theorem 7.2.1 tells us that the value is precisely the lower bound $\alpha(G)$ that we have established for $\vartheta(G)$ in Lemma 3.4.4.

On the other hand, we have also seen an *upper* bound for $\vartheta(G)$, namely $\vartheta(G) \leq \chi(\overline{G})$, the chromatic number of the complementary graph (this was the Sandwich Theorem 3.7.2). It turns out that this upper bound can be strengthened by a copositive *restriction* of the theta function program (3.7). Dukanovic and Rendl [DR10] show that if we add the constraint $Y + J_n \in \text{POS}_n$ to (3.7), where J_n is the $n \times n$ all-one matrix, we obtain a program whose value is the *fractional chromatic number* $\chi_f(\overline{G})$ of $\overline{G}$. It holds that $\chi_f(\overline{G}) \leq \chi(\overline{G})$, where strict inequality is possible.

The proof of Theorem 7.2.1 proceeds in two steps. First, we show that the value of (7.3) is at least $\alpha(G)$. For this, we compute the dual program (a maximization problem) and exhibit a feasible solution for it that has value $\alpha(G)$. Weak duality of cone programming then shows that the value of (7.3) is also at least $\alpha(G)$.

For the difficult direction (the value of (7.3) is at most $\alpha(G)$), we use the *Motzkin–Straus Theorem*, a beautiful result in graph theory that expresses the problem of computing $\alpha(G)$ as a (nonconvex) quadratic optimization problem over the unit simplex.

The Value Is at Least $\alpha(G)$

In Exercise 4.12, we have asked you to compute the semidefinite program dual to (3.7). We omit the proof of the following lemma, since it proceeds in precisely the same way, using the general definition of dual cone programs in Sect. 4.7, along with Lemma 4.5.3 (adjoint calculation). Not surprisingly, the dual of (3.7) is the relaxation of the dual of (7.3) obtained by replacing the constraint $Y \in \text{POS}_n$ with $Y \succeq 0$.

7.2.2 Lemma. *Let J_n denote the $n \times n$ all-one matrix. The cone program dual to the copositive program (7.3) is the copositive program*

$$
\begin{aligned}
\text{maximize} \quad & J_n \bullet X \\
\text{subject to} \quad & \text{Tr}(X) = 1 \\
& x_{ij} = 0, \ \text{if } \{i, j\} \in E \\
& X \in \text{POS}_n.
\end{aligned}
\tag{7.4}
$$

Now we are prepared for the "easy" part of the proof of Theorem 7.2.1.

7.2.3 Lemma. *The dual program (7.4) is feasible, and its value is at least $\alpha(G)$.*

The weak duality of cone programming (Theorem 4.7.2) now yields the following.

7.2.4 Corollary. *For every graph G, the value of the copositive program (7.3) is at least $\alpha(G)$.*

Proof of Lemma 7.2.3. We construct a feasible solution of value $\alpha(G)$. Let $\tilde{x} \in \mathbb{R}^n$ be the characteristic vector of a maximum independent set $I \subseteq V$, i.e., $\tilde{x}_i = 1$ if $i \in I$, and $\tilde{x}_i = 0$ otherwise. Now consider the matrix

$$\tilde{X} = \frac{1}{\alpha(G)}\tilde{x}\tilde{x}^T = \left(\frac{1}{\sqrt{\alpha(G)}}\tilde{x}\right)\left(\frac{1}{\sqrt{\alpha(G)}}\tilde{x}\right)^T \in \text{POS}_n.$$

We have

$$\text{Tr}(\tilde{X}) = \frac{1}{\alpha(G)}\sum_{i=1}^{n}\tilde{x}_i^2 = \frac{1}{\alpha(G)}\alpha(G) = 1,$$

and since I is an independent set, we have $\tilde{x}_{ij} = \frac{1}{\alpha(G)}\tilde{x}_i\tilde{x}_j = 0$ for all edges $\{i,j\}$. Hence, $\tilde{X}$ is feasible for (7.4), with value

$$J_n \bullet \tilde{X} = \sum_{i,j}\tilde{x}_{ij} = \frac{1}{\alpha(G)}\sum_{i,j}\tilde{x}_i\tilde{x}_j = \frac{1}{\alpha(G)}\sum_{i,j\in I}1 = \frac{\alpha(G)^2}{\alpha(G)} = \alpha(G). \quad \square$$

A Modified Program with the Same Value

Next we want to show that the value of program (7.3) is at most $\alpha(G)$. We start by rewriting it into a different program with the same value, setting the stage for the Motzkin–Straus theorem. Let $Y \in \text{SYM}_n$. The constraints

$$\begin{aligned} y_{ij} &= -1, \text{ if } \{i,j\} \in \bar{E} \\ y_{ii} &= t - 1, \text{ for all } i = 1, 2, \ldots, n \end{aligned}$$

can equivalently be expressed as

$$Y = tI_n + Z - J_n,$$

where Z is a matrix such that $z_{ij} = 0$ for $\{i,j\} \notin E$.

Let A_G be the adjacency matrix of G. If $Y = tI_n + Z - J_n$ is feasible for (7.3), and if z is the largest entry of Z, then $Y' = tI_n + zA_G - J_n$ is also feasible. Indeed, $Z' = zA_G$ also satisfies $z'_{ij} = 0$ for $\{i,j\} \notin E$, and since

$$Y' = \underbrace{Y}_{\in \text{COP}_n} + \underbrace{zA_G - Z}_{\geq 0},$$

we also have $Y' \in \text{COP}_n$ (recall that any nonnegative symmetric matrix is copositive).

Consequently, we may w.l.o.g. assume that Y in (7.3) is a matrix of the form $tI_n + zA_G - J_n$. This gives the following result.

7.2.5 Lemma. *The copositive program*

$$\begin{aligned} \text{Minimize} \quad & t \\ \text{subject to} \quad & tI_n + zA_G - J_n \in \text{COP}_n \\ & t, z \in \mathbb{R} \end{aligned} \qquad (7.5)$$

is feasible and has the same value as (7.3); in particular, the value is at least $\alpha(G)$.

The Motzkin–Straus Theorem

This theorem states that the problem of computing the independence number of a graph can be solved by minimizing a quadratic form over the unit simplex.

7.2.6 Theorem. *For every graph G,*

$$\frac{1}{\alpha(G)} = \min\{\mathbf{x}^T(A_G + I_n)\mathbf{x} \cdot \mathbf{x} \geq \mathbf{0}, \sum_{i=1}^{n} x_i - 1\}.$$

The original theorem is stated in a complementary setting and talks about cliques rather than independent sets; see Exercise 7.5. The version that we present here appears as Corollary 1 in [MS65].

Let us postpone the proof of the Motzkin–Straus theorem and first draw the conclusion.

The Value Is at Most $\alpha(G)$

7.2.7 Theorem. *For every graph G, the value of the copositive program (7.3) is at most $\alpha(G)$.*

Proof. We construct a feasible solution of value $\alpha(G)$ for the modified program (7.5). The statement then follows from Lemma 7.2.5.

If $\mathbf{x} \geq \mathbf{0}$ and $\sum_{i=1}^{n} x_i = 1$, the Motzkin–Straus theorem implies

$$\mathbf{x}^T(\alpha(G)(A_G + I_n))\mathbf{x} \geq 1 = \mathbf{x}^T J_n \mathbf{x},$$

meaning that

$$\mathbf{x}^T(\alpha(G)I_n + \alpha(G)A_G - J_n)\mathbf{x} \geq \mathbf{0}, \quad \text{if } \mathbf{x} \geq \mathbf{0}, \sum_{i=1}^{n} x_i = 1.$$

Now, since every nonzero $\mathbf{x}' \geq \mathbf{0}$ is a positive multiple of some $\mathbf{x} \geq 0$ with $\sum_{i=1}^{n} x_i = 1$, we actually have

$$\mathbf{x}^T(\alpha(G)I_n + \alpha(G)A_G - J_n)\mathbf{x} \geq \mathbf{0}, \quad \text{if } \mathbf{x} \geq 0.$$

This shows that the matrix $\tilde{Y} := \alpha(G)I_n + \alpha(G)A_G - J_n$ is copositive and hence a feasible solution of (7.5) with value $t = z = \alpha(G)$. $\qquad\square$

Proof of the Motzkin–Straus Theorem

It remains to prove the Motzkin–Straus theorem (Theorem 7.2.6). We define $f(\mathbf{x}) = \mathbf{x}^T(A_G + I_n)\mathbf{x}$ and let $m(G)$ denote the minimum of $f(\mathbf{x})$ over the unit simplex.

The inequality $m(G) \leq \frac{1}{\alpha(G)}$ is easy. If I is a maximum independent set with characteristic vector $\tilde{\mathbf{y}}$, then the vector $\tilde{\mathbf{x}} = \frac{1}{\alpha(G)}\tilde{\mathbf{y}}$ satisfies $\tilde{\mathbf{x}} \geq \mathbf{0}$, $\sum_{i=1}^{n} \tilde{x}_i = 1$ and thus

$$m(G) \leq f(\tilde{\mathbf{x}}) = 2 \sum_{\{i,j\} \in E} \tilde{x}_i \tilde{x}_j + \sum_{i \in I} \tilde{x}_i^2 = 0 + \frac{\alpha(G)}{\alpha(G)^2} = \frac{1}{\alpha(G)}.$$

For the other inequality, the strategy is as follows. We start from some minimizer $\mathbf{x}^*$ of f over the unit simplex and transform it into another minimizer $\mathbf{y}^*$ such that $J := \{i : y_i^* > 0\}$ is an independent set. As above, we then compute

$$m(G) = f(\mathbf{x}^*) = f(\mathbf{y}^*) = 2 \underbrace{\sum_{\{i,j\} \in E} y_i^* y_j^*}_{=0} + \sum_{i \in J}(y_i^*)^2 \geq \sum_{i \in J} \frac{1}{|J|^2} = \frac{1}{|J|} \geq \frac{1}{\alpha(G)}.$$

The second-to-last inequality uses the standard fact that $\sum_{i \in J} y_i^2$ is minimized subject to $\sum_{i \in J} y_i = 1$ if all the y_i are equal to $1/|J|$.

It remains to construct $\mathbf{y}^*$, given a minimizer $\mathbf{x}^*$. We define

$$F := \{\{i, j\} \in E : x_i^* x_j^* > 0\}.$$

If $F = \emptyset$, we set $\mathbf{y}^* := \mathbf{x}^*$ and are done. Otherwise, we choose $\{i, j\} \in F$ and define $\mathbf{z}$ through

$$z_k = \begin{cases} x_k^* + \varepsilon & \text{if } k = i \\ x_k^* - \varepsilon & \text{if } k = j \\ x_k^* & \text{otherwise,} \end{cases}$$

for some real number ε. A simple calculation shows that

$$f(\mathbf{z}) = f(\mathbf{x}^*) + \ell(\varepsilon),$$

where ℓ is a linear function in ε (due to $\{i, j\} \in E$, the quadratic term in ε vanishes). Since $\mathbf{z}$ is in the unit simplex for $|\varepsilon|$ sufficiently small, we must have $\ell(\varepsilon) = 0$, since otherwise, $f(\mathbf{z}) < f(\mathbf{x}^*)$ for some $\mathbf{z}$, contradicting our assumption that $\mathbf{x}^*$ is a minimizer of f.

We therefore have $f(\mathbf{z}) = f(\mathbf{x}^*)$, and choosing ε appropriately, we get $z_i = 0$. Replacing $\mathbf{x}^*$ with $\mathbf{z}$ yields a new minimizer for which the set F of "nonzero edges" has become strictly smaller. We can thus repeat the above process until F becomes empty and $\mathbf{y}^*$ is obtained.

7.3 Local Minimality Is coNP-hard

Theorem 7.2.1 immediately implies that copositive programming is NP-hard. In contrast to the case of semidefinite programming, we cannot even get a reasonable approximation of the optimal value in polynomial time, the reason being that such an approximation is already impossible for the independence number of a graph.

In this section, we want to derive some further hardness results, outside the (somewhat restricted) realm of copositive programming. They concern the problem of locally minimizing (or maximizing) a smooth function. Let $\mathcal{F}$ be a class of smooth functions $\phi \colon \mathbb{R}^n \to \mathbb{R}$ (by "smooth" we mean that ϕ has continuous partial derivatives of all orders). The problem $\mathrm{LocMin}(\mathcal{F})$ is to decide whether $\mathbf{0}$ (or equivalently, any other point) is a local minimum of a given function $\phi \in \mathcal{F}$ over $\mathbb{R}^n$. Our main result is the following.

7.3.1 Theorem. *The problem* $\mathrm{LocMin}(\mathcal{F})$ *is coNP-complete for the class* $\mathcal{F} := \{\phi_M : M \in \mathrm{SYM}_n\}$, *where the function* ϕ_M *is defined by*

$$\phi_M(\mathbf{x}) = (x_1^2, x_2^2, \dots, x_n^2) M (x_1^2, x_2^2, \dots, x_n^2)^T.$$

Here coNP is the class of "complements" of all problems in NP, i.e., problems for which the NO answer can be certified in polynomial time. In our case, the statement that $\mathrm{LocMin}(\mathcal{F})$ is coNP-complete means that the problem obtained from $\mathrm{LocMin}(\mathcal{F})$ by interchanging the YES and NO answers (i.e., we ask whether $\mathbf{0}$ is *not* a local minimum) is NP-complete.

This result (proved by Murty and Kabadi in 1987 [MK87]) is relevant if we want to understand to what extent nonlinear optimization problems can be solved in practice. A web search reveals numerous claims of the form that this and that code will find a local optimum of the nonlinear function at hand. Taking this for granted, the focus of the discussion is then rather on how this local optimum can be further improved to reach the global optimum in the end.

In view of Theorem 7.3.1, it is generally hard to even check whether a given solution is a local optimum; therefore, any claims of having reached local optimality must be taken with care. One misconception underlying such claims is that a local minimum is confused with a "pseudo-minimum," a point where the first derivative vanishes and the second derivative (Hessian) is positive semidefinite. Indeed, whether a solution is a pseudo-minimum can efficiently be checked in many cases, but even if it is, there is no guarantee that it actually is a true local minimum.

For example, every function considered in Theorem 7.3.1 has the point $\mathbf{0}$ as a pseudo-minimum. The difficulty is thus to distinguish pseudo-minima from true local minima.

To approach Theorem 7.3.1, we start with the following auxiliary result, which is also known [MK87], but for which the Motzkin–Straus theorem yields a simple proof.

7.3.2 Theorem. *Given an integer matrix $M \in \mathrm{SYM}_n$, it is coNP-complete to decide whether $M \in \mathrm{COP}_n$.*

Proof. The problem is in coNP, since there is a certificate for $M \notin \mathrm{COP}_n$ in form of $\mathbf{x} \in \mathbb{R}^n_+$ such that $\mathbf{x}^T M \mathbf{x} < 0$. It is not a priori clear (but true, see Exercise 7.6) that there is a certificate of size polynomial in the encoding size of M. To show that the problem is coNP-complete, we provide a reduction from the independent set problem: given a graph G and an integer k, does G have an independent set of size larger than k?

The reduction constructs the integer matrix $M(G, k) = kI_n + kA_G - J_n$. It is easy to check (see also the proof of Theorem 7.2.7) that $M(G, k)$ is copositive if and only if

$$\min\{\mathbf{x}^T(A_G + I_n)\mathbf{x} : \mathbf{x} \geq \mathbf{0}, \sum_{i=1}^n x_i = 1\} \geq \frac{1}{k}.$$

Hence

$$M(G, k) \notin \mathrm{COP}_n \quad \Leftrightarrow \quad k < \alpha(G),$$

by the Motzkin–Straus theorem. So G has an independent set of size larger than k if and only if $M(G, k)$ is not copositive.   □

As a corollary, we get our first "hardness of local minimality" result, for quadratic functions over the positive orthant.

7.3.3 Corollary. *Let $M \in \mathrm{SYM}_n$ be an integer matrix. It is coNP-complete to decide whether $\mathbf{0}$ is a local minimum of the quadratic form $\mathbf{x}^T M \mathbf{x}$ subject to $\mathbf{x} \geq \mathbf{0}$.*

Proof. As a direct consequence of Definition 7.1.1, $\mathbf{0}$ is a local minimum if and only if $M \in \mathrm{COP}_n$. Using coNP-completeness of the latter problem, the statement follows.   □

Now we are ready to prove the main result: hardness of local minimality in smooth unconstrained optimization.

Proof of Theorem 7.3.1. Let $\phi_M(\mathbf{x}) = (x_1^2, x_2^2, \ldots, x_n^2) M (x_1^2, x_2^2, \ldots, x_n^2)^T$. We show that $\mathbf{0}$ is a local minimum of ϕ if and only $\mathbf{0}$ is a local minimum of $\mathbf{x}^T M \mathbf{x}$ subject to $\mathbf{x} \geq \mathbf{0}$. Applying Corollary 7.3.3 then finishes the proof.

The point $\mathbf{0}$ is a local minimum of ϕ if and only if all $\mathbf{x}$ in some small neighborhood of $\mathbf{0}$ satisfy $\phi_M(\mathbf{x}) \geq 0$. Via the transformation $y_i = x_i^2$, this is in turn equivalent to the statement that all $\mathbf{y} \geq \mathbf{0}$ in some small neighborhood of $\mathbf{0}$ satisfy $\mathbf{y}^T M \mathbf{y} \geq 0$, meaning that $\mathbf{0}$ is a local minimum of $\mathbf{y}^T M \mathbf{y}$ subject to $\mathbf{y} \geq \mathbf{0}$. □

Exercises

7.1 Prove Lemma 7.1.5.

7.2 Characterize the triples $(x, y, z) \in \mathbb{R}^3$ for which the matrix

$$M = \begin{pmatrix} x & z \\ z & y \end{pmatrix} \in \text{SYM}_2$$

is (a) copositive, (b) completely positive. (Recall that M is positive semidefinite if and only if $x \geq 0, y \geq 0, xy \geq z^2$.)

7.3 Prove or disprove the following statements.

(i) If $M \in \text{SYM}_n$ is positive semidefinite, then the function $\mathbf{x} \mapsto \mathbf{x}^T M \mathbf{x}$ is convex over $\mathbb{R}^n$.

(ii) If $M \in \text{SYM}_n$ is copositive, then the function $\mathbf{x} \mapsto \mathbf{x}^T M \mathbf{x}$ is convex over $\mathbb{R}_+^n$.

(iii) If $M \in \text{SYM}_n$, $M \neq 0$ is completely positive, then the function $\mathbf{x} \mapsto \mathbf{x}^T M \mathbf{x}$ is strictly convex over $\mathbb{R}^n$.

A function $f: U \subseteq \mathbb{R}^n \to \mathbb{R}$ is convex over U if for all $\mathbf{x}, \mathbf{y} \in U$ and all $t \in (0, 1)$,

$$f((1 - t)\mathbf{x} + t\mathbf{y}) \leq (1 - t)f(\mathbf{x}) + tf(\mathbf{y}).$$

The function is strictly convex if the inequality is strict for $\mathbf{x} \neq \mathbf{y}$.

7.4 We know (Sect. 4.8) that for every matrix $C \in \text{SYM}_n$,

$$\max\{\mathbf{x}^T C \mathbf{x} : \|x\| = 1\} = \max\{C \bullet X : I_n \bullet X = 1, X \succeq 0\}.$$

This means that the problem of finding the largest eigenvalue of C reduces to a semidefinite program. Prove the following corresponding statement for copositive programming [BDDK$^+$00].

$$\max\{\mathbf{x}^T C \mathbf{x} : \textstyle\sum_{i=1}^n x_i = 1, \mathbf{x} \geq 0\} = \max\{C \bullet X : J_n \bullet X = 1, X \in \text{POS}_n\}.$$

7.5 Let $\omega(G)$ be the clique number of G, the size of the largest clique. Prove that

$$1 - \frac{1}{\omega(G)} = \max\{\mathbf{x}^T A_G \mathbf{x} : \mathbf{x} \geq \mathbf{0}, \textstyle\sum_{i=1}^n x_i = 1\}.$$

7.6 Let $M \in \text{SYM}_n$ be a matrix with integer coordinates. Prove that if there exists $\mathbf{x} \in \mathbb{R}_+^n$ such that $\mathbf{x}^T M \mathbf{x} < 0$, then there is also such an $\mathbf{x}$ whose binary encoding length is polynomial in the binary encoding length of M.

7.7 Prove that for the class of functions considered in Theorem 7.3.1, it is also coNP-complete to check whether the function is bounded from below over $\mathbb{R}^n$.

Part II
(by Jiří Matoušek)

Chapter 8
Lower Bounds for the Goemans–Williamson MaxCut Algorithm

8.1 Can One Get a Better Approximation Ratio?

8.1.1 The Goemans–Williamson algorithm for MaxCut is a showcase result in SDP-based approximation algorithms. As we know, the analysis yields the approximation ratio $\alpha_{GW} \approx 0.87856720578$. We can hardly avoid asking questions of the following kind: Is this strange-looking number just an artifact of the particular algorithm, or does it have any deeper significance for the MaxCut problem itself? Can we improve on the approximation ratio with some even more clever tricks?

In the last few hundred years we got used to steady progress of technology. In optimization, we had *linear programming*, and then *semidefinite programming*; so we may expect something even more powerful to come next.

8.1.2 So, can we improve on the GW algorithm's approximation ratio? People tried hard, and indeed there are positive results for various restricted cases:

- **YES** for *dense graphs* (cn^2 edges, $c > 0$ fixed). Here one can get a *polynomial-time approximation scheme* (PTAS), i.e., $(1-\varepsilon)$-approximation in polynomial time for every $\varepsilon > 0$ and $c > 0$ fixed.

  o This follows from an algorithmic version of the *Szemerédi regularity lemma* due to Alon et al. [ADRY94], and also from a framework of Arora et al. [AKK95]. (For faster versions, less astronomically depending on c, ε, see Frieze and Kannan [FK99] and Alon and Naor [AN96].)
  o The approach via the regularity lemma was generalized by Alon et al. [ACOH+10] to graphs that are not necessarily dense, but contain no "exceptionally dense spots" (i.e., subgraphs whose density exceeds the average density of the whole graph by more than a constant factor).

B. Gärtner and J. Matoušek, *Approximation Algorithms and Semidefinite Programming*, DOI 10.1007/978-3-642-22015-9_8,
© Springer-Verlag Berlin Heidelberg 2012

- **YES** for graphs of *bounded maximum degree* – see Feige et al. [FKL02].
- **YES** for graphs where the maximum cut is "large" or "small":

  o For graphs with *large* maximum cut (measured as a fraction of the number of edges), the analysis of the Goemans–Williamson algorithm yields a better approximation ratio, as was already observed in [GW95].
  o For graphs with *small* maximum cut, other rounding techniques also yield an improved ratio; see O'Donnell and Wu [OW08] for such a result and references.

This is all very nice, *but* can one get a better approximation ratio in general – for *all* input graphs?

8.1.3 Currently there are good reasons to believe that α_{GW} is the *worst-case optimal* approximation factor! Indeed, assuming the *Unique Games Conjecture* (UGC) and $P \neq NP$, there is *no* polynomial-time MaxCut algorithm with approximation ratio better than α_{GW}.

- We will mention the UGC and some of its consequences later. It is a somewhat technical-sounding statement, and the various hardness proofs based on it are quite demanding – they involve tools like PCP's, Fourier analysis, etc.

8.1.4 Without the UGC, assuming only $P \neq NP$, it is known that no polynomial-time algorithm can approximate MaxCut better than $\frac{16}{17} \approx 0.94$ (see Håstad [Hås01]; the proof uses advanced PCP techniques). Given a 3-Sat formula φ, Håstad's construction yields a graph $G = G(\varphi)$ such that

(1) MaxCut $\geq x$ if φ is satisfiable (where x can be efficiently computed from φ)
(2) MaxCut $\leq \frac{16}{17}x$ if φ is unsatisfiable.[1]

This must give "hard" instances for every MaxCut approximation algorithm, but it is not clear what they look like (essentially because the proof proceeds by contradiction; or put in another way, because we do not know what hard instances of 3-Sat look like).

8.1.5 We will not discuss these (difficult) relative results here. Rather we will cover *absolute* lower bounds for the Goemans–Williamson algorithm itself.

- These are hands-on "hard" instances, but *only* for this particular algorithm (or class of algorithms).
- Beautiful mathematics is involved, and one can learn useful tricks.
- We will not cover the strongest known results. Rather, we want to introduce interesting methods on somewhat weaker, but simpler results.

---

[1] Actually, $(\frac{16}{17} + \varepsilon)x$, where $G(\varphi)$ also depends on ε.

8.2 Approximation Ratio and Integrality Gap

8.2.1 Let $G = (V, E)$ be an input graph, $V = \{1, 2, \ldots, n\}$.

- As we recall from Chap. 1, MAXCUT for G can be written as the following integer quadratic program:

$$\max\left\{ \sum_{\{i,j\}\in E} \frac{1 - x_i x_j}{2} : x_1, \ldots, x_n \in \{\pm 1\} \right\}.$$

The optimum, denoted by Opt or Opt(G), is the true size of a maximum cut.

- The *semidefinite relaxation* of MAXCUT, written as a *vector program*, is

$$\max\left\{ \sum_{\{i,j\}\in E} \frac{1 - \mathbf{v}_i^T \mathbf{v}_j}{2} : \|\mathbf{v}_1\| = \cdots = \|\mathbf{v}_n\| = 1 \right\}; \qquad \text{(GW)}$$

we denote its optimum value by SDP or SDP(G).

- We let Algo = Algo(G) denote the expected size of the cut found by the random hyperplane rounding in the GW algorithm.[2] Thus, the approximation ratio of the Goemans–Williamson algorithm can be written as

$$\inf_{G} \frac{\text{Algo}(G)}{\text{Opt}(G)}.$$

Here is a pictorial summary of the considered quantities:

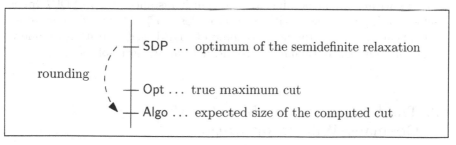

rounding

SDP ... optimum of the semidefinite relaxation

Opt ... true maximum cut

Algo ... expected size of the computed cut

8.2.2 A perhaps less intuitive but very important notion is the *integrality gap* (of the considered semidefinite relaxation):

$$\text{Gap} := \sup_{G} \frac{\text{SDP}(G)}{\text{Opt}(G)}.$$

---

[2] Actually, we can assume that the algorithm *always* finds a cut with at least Algo edges. For example, we can run it repeatedly until it succeeds. There is also a deterministic (derandomized) version by Mahajan and Ramesh [MR99]. The derandomization is not easy, and in particular, the derandomized version in the original GW paper is not correct.

- The integrality gap is a general notion in the theory of integer programming. It measures how far the optimum of a semidefinite (or linear) relaxation is from the optimum of the underlying integer program.
- A somewhat unpleasant feature of the above definitions is that the approximation ratio is a number *smaller* than 1, while the integrality gap is *larger* than 1.

  ○ However, one probably should not try to change the definition of the approximation ratio since it is too well established. The definition of the integrality gap as above is also quite common. Moreover, if we defined integrality gap the other way round, as $\inf(\text{Opt}/\text{SDP})$, then a *smaller number* would mean a *larger (more significant) gap*, which would sound very strange.

8.2.3 The integrality gap, or more precisely, the quantity $1/\text{Gap}$, is usually an upper bound on the approximation ratio of (maximization) algorithms based on semidefinite programming. (The same applies to algorithms based on linear programming.)

- This should not be taken as a precise mathematical statement, but rather as a rule of thumb. We have not specified the class of the considered algorithms. So, for example, the algorithm might "cheat": first solve some semidefinite relaxation, say with a significant integrality gap, and then compute a very good approximate solution of the original problem by some completely different method.
- Why do we expect approximation ratio no better than $1/\text{Gap}$? Think of a hypothetical algorithm that beats $1/\text{Gap}$ on all input instances. Such an algorithm would be able to distinguish instances where SDP/Opt is close to Gap from those with SDP/Opt close to 1. This task seems to go significantly beyond the usual scheme of SDP-based algorithms, namely, compute an SDP optimum and round it to an integral solution.

8.3 The Integrality Gap Matches the Goemans–Williamson Ratio

8.3.1 For the Goemans–Williamson algorithm, the analysis in Chap. 1 shows that $\text{Algo}/\text{SDP} \geq \alpha_{\text{GW}}$, and since $\text{Opt} \geq \text{Algo}$, we have $\text{Gap} = \sup(\text{SDP}/\text{Opt}) \leq 1/\alpha_{\text{GW}}$. The next theorem shows that the integrality gap actually equals $1/\alpha_{\text{GW}}$.

- The proof will require a substantial effort, and it will touch upon several interesting geometric phenomena.
- Nowadays, instances with large integrality gap are usually constructed by newer techniques, such as PCP-related technology (using dictatorship tests, long codes, etc.).

> **8.3.2 Theorem** (Feige and Schechtman [FS02]). *The integrality gap of the Goemans–Williamson semidefinite relaxation of* MAXCUT *satisfies* Gap $\geq \frac{1}{\alpha_{GW}} \approx 1.1382$. *In other words, for every* $\varepsilon > 0$ *there exists a graph* G *with*
> $$\frac{\text{SDP}}{\text{Opt}} \geq \frac{1}{\alpha_{GW}} - \varepsilon.$$

8.3.3 To bound the integrality gap from below, we need to exhibit a graph, provide a *feasible* solution of the SDP (typically the easier part), and show that there is no large cut (the harder part).

- Warm-up: For $G := C_5$, we have Opt $= 4$, and the feasible SDP solution in the next picture shows that SDP $\geq 5(1 - \cos\frac{4\pi}{5})/2 \approx 4.5225$:

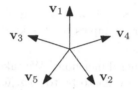

(this is actually an *optimal* SDP solution; see Exercise 8.2). Hence Gap $\geq$ 1.1305.

- The theorem asserts that Gap $\geq 1/\alpha_{GW} \approx 1.1382$, and we already have Gap ≥ 1.1305. So all the fuss below is about 0.7% of the value of Gap. (But really about *understanding*.)

8.3.4 We recall the Goemans–Williamson analysis. For an edge $\{i, j\}$ represented by vectors $\mathbf{v}_i, \mathbf{v}_j$ with angle ϑ, we have:

- Contribution $\frac{1 - \mathbf{v}_i^T \mathbf{v}_j}{2} = \frac{1 - \cos\vartheta}{2}$ to SDP.
- Expected contribution $\frac{\vartheta}{\pi}$ to Algo $\leq$ Opt.

The ratio of these contributions is $\frac{\pi}{2}\frac{1 - \cos\vartheta}{\vartheta}$, and $1/\alpha_{GW}$ is the *maximum* possible value of this ratio, attained for $\vartheta = \vartheta_{GW} \approx 133.563°$. Thus, if SDP/Opt should be close to $1/\alpha_{GW}$, we need an SDP solution such that the angle $\angle\mathbf{v}_i\mathbf{v}_j$ is close to ϑ_{GW} for most of the edges $\{i, j\} \in E(G)$.

8.3.5 A simple proof idea: Place a large finite set of points densely on the unit sphere S^{d-1}. They are the vertices of G and, at the same time, they define a feasible SDP solution since they are unit vectors. The edges should correspond to pairs of vectors whose angle is close to ϑ_{GW}. This looks promising since

- SDP is at least about $|E|\frac{1 - \cos\vartheta_{GW}}{2}$.
- The expected size of a random hyperplane cut is close to $|E|\frac{\vartheta_{GW}}{\pi}$.

8.3.6 *But*, it is not enough to look at hyperplane cuts! (We want an upper bound on Opt, not only on Algo.) For example, for $d = 2$, let us consider

8 equally spaced unit vectors in the plane. The edges correspond to angles 135°, instead of 133.563:

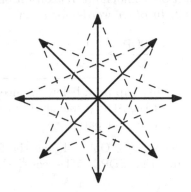

The resulting graph is an 8-cycle and so Opt = 8, but all hyperplane cuts have only 6 edges. Planar examples of this kind do not work for getting a better integrality gap...

8.3.7 Remedy: Go to higher dimensions! We take $d = d(\varepsilon)$ sufficiently large, depending on how close we want to be to $1/\alpha_{\mathrm{GW}}$.

8.3.8 The proof of Theorem 8.3.2 has two main parts:

- **The continuous analog.** We define an *infinite* "continuous" graph G_c with vertex set S^{d-1} (all unit vectors in $\mathbb{R}^d$) and edges connecting unit vectors whose angle is within $\vartheta_{\mathrm{GW}} \pm \delta$, for a suitable small $\delta = \delta(\varepsilon)$. We prove geometrically that "the largest cuts are hyperplane cuts." (We cannot *count* the edges of a cut, since cuts are infinite – we have to introduce a *measure* on the edge set.)
- **Discretization.** We divide S^{d-1} into a large but finite number of small-diameter pieces (cells), and in each cell we choose a single vertex of the "discrete graph" G. These vertices are again connected if their angle is within $\vartheta_{\mathrm{GW}} \pm \delta$. We argue that if G had a large cut, then G_c also has a large cut.

The Continuous Graph, Caps, and Isoperimetry

8.3.9 Let μ be the *uniform probability measure on* S^{d-1}. Here is a "pedestrian" definition: For $A \subseteq S^{d-1}$, we set

$$\mu(A) := \frac{\lambda^d(\tilde{A})}{\lambda^d(B^d)},$$

where

- $\lambda^d(.)$ is the d-dimensional volume in $\mathbb{R}^d$ (actually, the Lebesgue measure).

- $\tilde{A} := \bigcup_{a \in A} \mathbf{0}\mathbf{a}$ is the union of all segments connecting points of A to the center of the sphere.
- B^d is the d-dimensional unit ball.

Here is a 3-dimensional illustration:

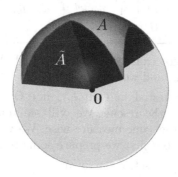

8.3.10 Now μ^2 measures subsets of $S^{d-1} \times S^{d-1}$, such as cuts in our continuous graph G_c.

- Formally, μ^2 can be defined as a *product measure*, which is a general construction in measure theory (generalizing the passage from $\lambda^1(.)$ to $\lambda^2(.)$).
- We could also proceed in a way similar to the above definition of μ using $\tilde{A}$.
- In any case, we will use only very simple and intuitive properties of μ^2.

8.3.11 We fix $\delta = \delta(\varepsilon) > 0$ sufficiently small, and we define the *continuous graph* $G_c = (S^{d-1}, E_c)$, where

$$E_c := \left\{ (\mathbf{x}, \mathbf{y}) \in S^{d-1} \times S^{d-1} : \angle \mathbf{x}\mathbf{y} \in [\vartheta_{\mathrm{GW}} - \delta, \vartheta_{\mathrm{GW}} + \delta] \right\}$$

- Here $\angle \mathbf{x}\mathbf{y}$ denotes the angle of vectors $\mathbf{x}$ and $\mathbf{y}$. It lies in the interval $[0, \pi]$, and it can be expressed as $\arccos \frac{\mathbf{x}^T \mathbf{y}}{\|\mathbf{x}\| \cdot \|\mathbf{y}\|}$.
- For convenience, the edges are *ordered* pairs $(\mathbf{x}, \mathbf{y})$.

8.3.12 What is a *cut* in G_c? For a measurable $A \subset S^{d-1}$,

$$\mathrm{cut}(E_c, A) := \left\{ (\mathbf{x}, \mathbf{y}) \in E_c : \text{ exactly one of } \mathbf{x}, \mathbf{y} \text{ lies in } A \right\}.$$

We will measure the size of cuts in G_c as the fraction of the "edges" of E_c contained in the cut. In particular,

$$\mathrm{Opt}(G_c) := \sup_A \frac{\mu^2(\mathrm{cut}(E_c, A))}{\mu^2(E_c)}.$$

8.3.13 What is the worst A (largest cut)? This is not easy to tell! But here is the key trick: We ignore the upper bound on $\angle \mathbf{x}\mathbf{y}$; we consider the *filled continuous graph* $G_c^+ = (S^{d-1}, E_c^+)$, where

$$E_c^+ := \left\{ (\mathbf{x}, \mathbf{y}) \in S^{d-1} \times S^{d-1} : \angle \mathbf{xy} \geq \vartheta_{\mathrm{GW}} - \delta \right\}.$$

Unlike for G_c, one *can* describe maximum cuts in G_c^+!

8.3.14 Proposition. Opt(G_c^+) *is attained by hyperplane cuts. That is, for every (measurable)* $A \subseteq S^{d-1}$ *we have* $\mu^2(\mathrm{cut}(E_c^+, A)) \leq \mu^2(\mathrm{cut}(E_c^+, H))$*, where H is a hemisphere.*

This proposition is not exactly easy to prove. It belongs to the area of *isoperimetric inequalities* and we will discuss it later.

8.3.15 But we are interested in Opt(G_c), not Opt(G_c^+). Here the high-dimensionality comes to the rescue: We will see that for $d = d(\delta)$ large, G_c and G_c^+ are almost the same measure-wise. We will see this in the proof of Lemma 8.3.19 below, but first we prepare some tools.

8.3.16 A *spherical cap* is a set of the form

$$C = \{ \mathbf{x} \in S^{d-1} : \angle \mathbf{xv} \leq \beta \}$$

for some $\mathbf{v} \in S^{d-1}$ (the *center* of C) and β (the *angle* of C). We write C_β for (some) cap with angle β.

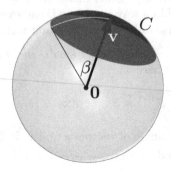

8.3.17 Claim. (Contrary to low-dimensional intuition.) *For every $\delta, \eta > 0$ and $\beta \in [0, \frac{\pi}{2} - \delta]$ there is a dimension d in which $\mu(C_\beta) \leq \eta \cdot \mu(C_{\beta + \delta})$.*

Thus, in high dimensions, almost all of the measure of a spherical cap lies very close to the boundary. (Keyword: *measure concentration*.)

8.3.18 Proof:

- The set $\tilde{C}_\beta$ (the cone over C_β) is contained in the cylinder of height 1 and radius $r := \sin \beta$.

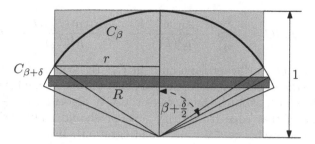

- $\tilde{C}_{\beta+\delta}$ contains the cylinder of radius $R := \sin(\beta + \frac{\delta}{2})$ and height $h = \cos(\beta + \frac{\delta}{2}) - \cos(\beta + \delta)$.
- The ratio of volumes of these cylinders is $\frac{1}{h}(\frac{r}{R})^{d-1} < \eta$ for d large enough. This verifies the claim. $\qquad\square$

Let us remark that our calculation in this proof was very rough, but fairly precise quantitative bounds on $\mu(C_\beta)$ are known.

8.3.19 The next lemma uses the above tools for showing that no cut in G_c is much larger than hyperplane cuts.

Lemma. *We have*

$$\mathrm{Opt}(G_c) \leq \frac{\vartheta_{\mathrm{GW}}}{\pi} + O(\delta).$$

8.3.20 Proof:

- We have $\mu^2(E_c^+) = \mu(C_{\pi-\vartheta_{\mathrm{GW}}+\delta})$, since the neighborhood of each $\mathbf{x}$ in G_c^+ is the cap with center $-\mathbf{x}$ and angle $\pi - \vartheta_{\mathrm{GW}} + \delta$. Similarly, $\mu^2(E_c) = \mu(C_{\pi-\vartheta_{\mathrm{GW}}+\delta}) - \mu(C_{\pi-\vartheta_{\mathrm{GW}}-\delta})$.
- Using Claim 8.3.17 with 2δ instead of δ and with $\eta := \delta$, we find

$$\begin{aligned}
\mu^2(E_c) &= \mu(C_{\pi-\vartheta_{\mathrm{GW}}+\delta}) - \mu(C_{\pi-\vartheta_{\mathrm{GW}}-\delta}) \\
&\geq \mu(C_{\pi-\vartheta_{\mathrm{GW}}+\delta})(1-\delta) = (1-\delta)\mu^2(E_c^+).
\end{aligned}$$

So E_c^+ is just a bit larger than E_c.
- We have

$$\sup_A \mu^2(\mathrm{cut}(E_c, A)) \leq \sup_A \mu^2(\mathrm{cut}(E_c^+, A)) \leq \mu^2(\mathrm{cut}(E_c^+, H)),$$

where H is a hemisphere (Proposition 8.3.14).
- Then we use that $E_c^+ \setminus E_c$ is small:

$$\begin{aligned}
\mu^2(\mathrm{cut}(E_c^+, H)) &\leq \mu^2(\mathrm{cut}(E_c, H)) + \mu^2(E_c^+ \setminus E_c) \\
&= \mu^2(\mathrm{cut}(E_c, H)) + \mu^2(E_c^+) - \mu^2(E_c) \\
&\leq \mu^2(\mathrm{cut}(E_c, H)) + \frac{\delta}{1-\delta}\mu^2(E_c).
\end{aligned}$$

- So

$$\text{Opt}(G_c) = \sup_A \frac{\mu^2(\text{cut}(E_c, A))}{\mu^2(E_c)} \leq \frac{\mu^2(\text{cut}(E_c, H))}{\mu^2(E_c)} + O(\delta).$$

- Now H defines a "hyperplane cut" in G_c. By symmetry all hyperplane cuts have the same size. The expected contribution of an edge to a random hyperplane cut, as in the analysis of the GW algorithm, is $\frac{\vartheta}{\pi}$, where ϑ is the angle of that edge (to be rigorous, we should really write a double integral and exchange the order of integration). In G_c the angles of all edges are within $\vartheta_{\text{GW}} \pm \delta$, and the lemma follows. $\square$

8.3.21 So we know that the continuous graph G_c has SDP/Opt arbitrarily close to $1/\alpha_{\text{GW}}$ (well, except that we have not proved Proposition 8.3.14). Next, we want to "discretize" G_c.

From the Continuous Graph to the Discrete Graph

8.3.22 To discretize, we need an "obvious-looking" result.

> **Lemma.** For every d and every $\gamma > 0$ there exists an integer n such that S^{d-1} can be subdivided into cells $U_1, \ldots, U_n$ with $\mu(U_i) = \frac{1}{n}$ and $\text{diam}(U_i) \leq \gamma$ for all i.

8.3.23 Proof (I do not know of a *really* simple proof; this one also gives essentially the best known quantitative bound on n in terms of d and γ):

- Avoiding some fractions, we actually go for $\text{diam}(U_i) \leq O(\gamma)$.
- Let $P = \{\mathbf{p}_1, \ldots, \mathbf{p}_k\} \subset S^{d-1}$ be an inclusion-maximal γ-*separated set* (which means that every two points in P have distance at least γ). Thus, every $\mathbf{x} \in S^{d-1}$ has some point of P at distance at most γ.
- Let V_i be the *Voronoi cell* of $\mathbf{p}_i$ in S^{d-1} (the set of all $\mathbf{x} \in S^{d-1}$ closer to $\mathbf{p}_i$ than to any other $\mathbf{p}_j$).

  ○ V_i contains the $\frac{\gamma}{2}$-ball around $\mathbf{p}_i$, and is contained in the γ-ball (balls taken within S^{d-1}).
  ○ So the V_i have small diameter and their measures vary only within a factor depending only on d (about 2^d). But we want *exactly* the same measures.

- Idea: Choose n so that $\frac{1}{n} < \min_i \mu(V_i)$. Let each vertex exchange some land with its neighbors so that everyone's territory is an integer multiple of $\frac{1}{n}$. Then we just slice each of the resulting territories into regions of measure $\frac{1}{n}$.

- We need to organize the territory exchange. Call i, j *neighbors* if $\|\mathbf{p}_i - \mathbf{p}_j\| \leq 2\gamma$. The graph on P with edges corresponding to neighbors is connected.

  ○ Why? Consider an arc α from some $\mathbf{p}_i$ to some $\mathbf{p}_j$; each point of α has some $\mathbf{p}_k$ at most γ away, so instead of following α, jump through these $\mathbf{p}_k$. Jump length 2γ suffices.

- Choose a rooted spanning tree T of this neighbor graph (so we can speak of the root, the leaves, and father-son relations among the vertices).
- In the first step of the territory exchange, each leaf $\mathbf{p}_i$ in T "rounds down" its territory to the nearest multiple of $\frac{1}{n}$: It gives a portion $D_i \subset V_i$ of measure $\mu(V_i) - \frac{1}{n}\lfloor n\mu(V_i)\rfloor$ to its father, so its remaining territory is $W_i := V_i \setminus D_i$. After that, such a leaf $\mathbf{p}_i$ is *finished*.
- The exchange procedure continues in a similar manner towards the root. Let us say that a non-leaf vertex $\mathbf{p}_i$ is *ready* if all of its sons are finished. For such $\mathbf{p}_i$, we let $V_i' := V_i \cup \bigcup_{\mathbf{p}_j \text{ son of } \mathbf{p}_i} D_j$ be the original Voronoi region of $\mathbf{p}_i$ plus the territory received from its sons.

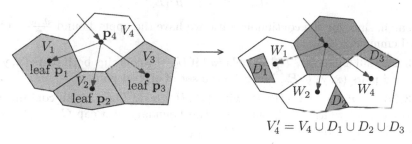

$$V_4' = V_4 \cup D_1 \cup D_2 \cup D_3$$

- As soon as some non-root vertex $\mathbf{p}_i$ becomes ready, it "rounds down" its territory: It gives a piece $D_i \subseteq V_i$ of measure $\mu(D_i) = \mu(V_i') - \frac{1}{n}\lfloor n\mu(V_i')\rfloor$ to its father. The remaining territory $V_i' \setminus D_i$ is called W_i.

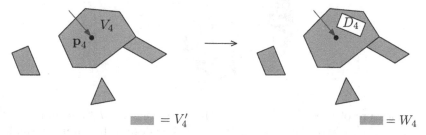

$$\blacksquare = V_4' \qquad \blacksquare = W_4$$

- In this way, we obtain W_i, with $\mu(W_i)$ a positive integer multiple of $\frac{1}{n}$, for each $\mathbf{p}_i$ except for the root. For the root $\mathbf{p}_r$, we can set $W_r := V_r'$, since $\mu(V_r') = 1 - \sum_{i \neq r} \mu(W_i)$ is also a positive integer multiple of $\frac{1}{n}$.
- We have $\mathrm{diam}(W_i) = O(\gamma)$, since $\mathbf{p}_i$ received territory only from the original Voronoi regions of its neighbors.
- Finally, we chop every W_i into $n\mu(W_i)$ pieces, of measure $\frac{1}{n}$ each, arbitrarily. The lemma is proved. $\qquad\square$

8.3.24 Now we are going to continue with the proof of Theorem 8.3.2. Based on the continuous graph G_c, we define the *discrete graph* G. Take γ very small compared to δ and divide S^{d-1} into n cells as in Lemma 8.3.22. Pick a vertex $\mathbf{v}_i$ in every cell U_i and set

$$V(G) \; := \; \{\mathbf{v}_1, \ldots, \mathbf{v}_n\},$$
$$E \; = \; E(G) := \{\{\mathbf{v}_i, \mathbf{v}_j\} : \angle \mathbf{v}_i \mathbf{v}_j \in [\vartheta_{\mathrm{GW}} - \delta, \vartheta_{\mathrm{GW}} + \delta]\}.$$

8.3.25 Clearly,

$$\frac{\mathsf{SDP}(G)}{|E|} \geq \frac{1 - \cos \vartheta_{\mathrm{GW}}}{2} + \varepsilon$$

for δ sufficiently small. It remains to bound $\mathsf{Opt}(G)$.

8.3.26 Let $A \subset V(G)$ define a cut with edge set $\mathrm{cut}(E, A)$. Set $A_c := \bigcup_{\mathbf{v}_i \in A} U_i$. We want to show that the difference of the "discrete ratio" and "continuous ratio"

$$\frac{|\mathrm{cut}(E, A)|}{|E|} - \frac{\mu^2(\mathrm{cut}(E_c, A_c))}{\mu^2(E_c)}$$

is small, since for the continuous ratio we have the upper bound $\frac{\vartheta_{\mathrm{GW}}}{\pi} + O(\delta)$ by Lemma 8.3.19.

8.3.27 Call a pair $\{U_i, U_j\}$ of cells *bad* if $U_i \times U_j$ contains both pairs $(\mathbf{x}, \mathbf{y}) \in E_c$ and pairs $(\mathbf{x}, \mathbf{y}) \notin E_c$. Let $\mathcal{B}$ be the set of all bad pairs.

8.3.28 Fix i; where are the U_j with $\{U_i, U_j\}$ bad? They are all contained in two bands of width $O(\gamma)$, one along the boundary of a cap $C_{\pi - \vartheta_{\mathrm{GW}} - \delta}$ and the other along the boundary of $C_{\pi - \vartheta_{\mathrm{GW}} + \delta}$.

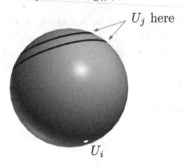

The measure of these bands is $O(\gamma)$ (here we consider d *fixed* and $\gamma \to 0$). So $|\mathcal{B}| \leq \beta n^2$, where β can be made as small as desired by taking γ small.

8.3.29 Each cut edge $\{\mathbf{v}_i, \mathbf{v}_j\} \in \mathrm{cut}(E, A)$ "contributes" the sets $U_i \times U_j$ and $U_j \times U_i$, each of measure n^{-2}, to the continuous cut $\mathrm{cut}(E_c, A_c)$, *unless* $\{U_i, U_j\}$ is a bad pair. So

$$\mu^2(\mathrm{cut}(E_c, A_c)) \;\geq\; 2n^{-2}(|\mathrm{cut}(E, A)| - |\mathcal{B}|)$$
$$\geq\; 2n^{-2}|\mathrm{cut}(E, A)| - 2\beta.$$

By a similar logic, $\mu^2(E_c) \leq 2n^{-2}|E| + 2\beta$.

8.3.30 Now for all $\gamma < \gamma_0$, with γ_0 suitably small, we have $2n^{-2}|E| \geq \mu^2(E_c) - 2\beta \geq \frac{1}{2}\mu^2(E_c)$. This is bounded away from 0 by some function of d and δ, independent of γ. So we can take γ small enough and make sure that

$$\frac{|\text{cut}(E, A)|}{|E|} \leq \frac{\mu^2(\text{cut}(E_c, A_c))}{\mu^2(E_c)} + \delta.$$

8.3.31 It follows that $\text{Opt}(G) \leq \frac{\vartheta_{\text{GW}}}{\pi} + O(\delta)$, and this shows that the integrality gap for G is at least $\alpha_{\text{GW}}^{-1} - \varepsilon$, as Theorem 8.3.2 claims. $\square$

The Isoperimetric Inequality: Proof Sketch

8.3.32 The mother of all isoperimetric inequalities asserts that *among all planar geometric figures of a given perimeter, the circular disk has the largest possible area.* We will first talk about this one, and then see how the proof idea generalizes.

8.3.33 Here is a technically much more convenient formulation:

Theorem (A planar isoperimetric inequality). *Let $t > 0$. If $A \subset \mathbb{R}^2$ is a compact set and C is the circular disk of the same area, then $\lambda^2(C_t) \leq \lambda^2(A_t)$. Here A_t stands for the t-neighborhood of A, consisting of all points with distance at most t to A, and $\lambda^2(.)$ is the area.*

An illustration follows:

The theorem claims that if the dark areas are the same, then the light gray area is the smallest for a disk.

- The statement above with perimeter can be obtained by considering the limit for $t \to 0$.

8.3.34 Isoperimetric inequalities constitute an important and advanced area, with many techniques and results. One way of proving such inequalities is *symmetrization*. There are various symmetrizing operations; we discuss one of them, which we call "foldup."

- If A is a compact planar set and H is a halfplane, then $\text{foldup}_H(A)$ is the set obtained by replacing each point $\mathbf{x} \in A \setminus H$ with its mirror reflection $\sigma_H \mathbf{x}$

w.r.t. the boundary of H, *provided* that $\sigma_H \mathbf{x} \notin A$. Formally, $\mathrm{foldup}_H(A) = (A \cap H) \cup (\sigma_H(A \setminus H)) \cup \{\mathbf{x} \in A \setminus H : \sigma_H \mathbf{x} \in A\}$.

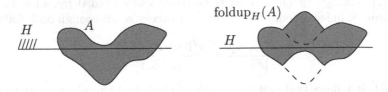

8.3.35 Lemma (Properties of foldup). *Let A be compact. (Measurable would be enough.) Then we have:*

(i) $\lambda^2(\mathrm{foldup}_H(A)) = \lambda^2(A)$.
(ii) *If C is a (circular) disk and H contains its center, then $\mathrm{foldup}_H(C) = C$.*
(iii) $\lambda^2(\mathrm{foldup}_H(A)_t) \le \lambda^2(A_t)$, *for every $t > 0$.*
(iv) $\mathrm{foldup}_H(A \cap B) \subseteq \mathrm{foldup}_H(A) \cap \mathrm{foldup}_H(B)$.

8.3.36 Proof:

- (i) and (ii) are obvious. (Actually, if (i) is to be done properly, we need a bit of measure theory.)
- For (iii), we prove $\mathrm{foldup}_H(A)_t \subseteq \mathrm{foldup}_H(A_t)$ and then use (i). The inclusion is left as an exercise.
- Part (iv) is routine and we omit it.

8.3.37 Outline of a proof of Theorem 8.3.33.

- Let $t > 0$ be fixed. Let C be a disk of area 1 with center $\mathbf{c}$. Among all planar compact sets A of area 1, let us consider those with $\lambda^2(A_t)$ minimum possible, and among these, let B have the largest overlap with C; i.e., B maximizes $\lambda^2(B \cap C)$. We want to prove $B = C$.
- For contradiction, suppose that $B \ne C$. Then $\lambda^2(B \setminus C) = \lambda^2(C \setminus B) > 0$. (If $\lambda^2(C \setminus B) = 0$, then $C \setminus B = \emptyset$ since $C \setminus B$ is open as a subset of C; therefore, $C \subseteq B$, and if B had a point outside C, we would get $\lambda^2(B_t) > \lambda^2(C_t)$.)
- Since $\lambda^2(C \setminus B) > 0$, there is $\mathbf{p} \in C \setminus B$ such that for every sufficiently small disk $D_\mathbf{p}$ centered at $\mathbf{p}$, the set $C \setminus B$ fills at least 99% of the area in $D_\mathbf{p}$ (by the *Lebesgue density theorem*[3]). Similarly, there is $\mathbf{q} \in B \setminus C$ such that for every sufficiently small disk $D_\mathbf{q}$ centered at $\mathbf{q}$, the set $B \setminus C$ fills at least 99% of the area of $D_\mathbf{p}$. Let us fix such disks $D_\mathbf{p}$ and $D_\mathbf{q}$ of equal radii.

---

[3] If $A \subseteq \mathbb{R}^d$ is measurable, then for all $\mathbf{x} \in A$, except for a set of measure 0, the *density* $d_A(\mathbf{x}) := \lim_{\varepsilon \to 0} \lambda^d(A \cap B(\mathbf{x}, \varepsilon)) / \lambda^d(B(\mathbf{x}, \varepsilon))$ (exists and) equals 1. Here $B(\mathbf{x}, \varepsilon)$ is the ball of radius ε centered at $\mathbf{x}$.

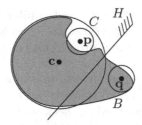

- Let H be the halfplane whose boundary bisects the segment $\mathbf{pq}$ and that contains $\mathbf{c}$. Since $\mathbf{p} \in C$ and $\mathbf{q} \notin C$, $\mathbf{c}$ is closer to $\mathbf{p}$ than to $\mathbf{q}$, and so $\mathbf{p} \in H$ and $\mathbf{q} \notin H$.
- Set $B^* := \mathrm{foldup}_H(B)$. We claim $\lambda^2(B^* \cap C) > \lambda^2(B \cap C)$, which will be a contradiction.
- Using Lemma 8.3.35(ii) and then (iv), we have $B^* \cap C = \mathrm{foldup}_H(B) \cap \mathrm{foldup}_H(C) \supseteq \mathrm{foldup}_H(B \cap C)$. We claim that the inclusion is strict in terms of measure, i.e., $\lambda^2(B^* \cap C) > \lambda^2(\mathrm{foldup}_H(B \cap C))$:

  - Most of $D_\mathbf{q}$ is contained in B, and since $D_\mathbf{q}$ "gets folded" to $D_\mathbf{p}$, we get that most of $D_\mathbf{p}$ is contained in $B^* \cap C$.
  - On the other hand, most of $D_\mathbf{p}$ is disjoint from B, and since C is disjoint from $D_\mathbf{q}$, we have $D_\mathbf{p} \cap \mathrm{foldup}_H(B \cap C) = D_\mathbf{p} \cap B$. Thus, most of $D_\mathbf{p}$ is disjoint from $\mathrm{foldup}_H(B \cap C)$, and so $\lambda^2(B^* \cap C) > \lambda^2(\mathrm{foldup}_H(B \cap C)) = \lambda^2(B \cap C)$.

- So B^* has a strictly larger overlap with C than B, which contradicts our initial choice of B.
- This was **not** a full proof! Why? Because the *existence* of a minimizer B is far from clear. (This was the gap in several early "proofs" of the planar isoperimetric inequality.) A not completely trivial topological argument is needed, which we omit (see, e.g., Feige and Schechtman [FS02] for an argument of this kind).

This is not the simplest proof of the planar isoperimetric inequality (there are several truly beautiful proofs known), but it generalizes to the setting we are interested in.

8.3.38 Now we work towards the isoperimetric inequality we wanted – Proposition 8.3.14. By the just indicated method one proves the following:

Lemma. Let $\vartheta \in (0, \pi)$, and let $E^{(\vartheta)} := \{(\mathbf{x}, \mathbf{y}) \in S^{d-1} \times S^{d-1} : \angle \mathbf{xy} \geq \vartheta\}$. Let $a \in [0, 1]$. Then among all closed sets $A \subseteq S^{d-1}$ with $\mu(A) = a$, the quantity $\mu^2(\mathrm{cut}(E^{(\vartheta)}, A))$ is maximized by a spherical cap.

8.3.39 Sketch of proof:

- Among all closed $A \subseteq S^{d-1}$ with $\mu(A) = a$, we consider those maximizing $\mu^2(\mathrm{cut}(E^{(\vartheta)}, A))$, and among those, pick B maximizing $\mu(B \cap C)$, where C is a fixed spherical cap of measure a. Again, the existence is nontrivial (and omitted – see the Feige–Schechtman paper).

- We want to show that $B = C$, and similar to the previous proof, for this it suffices to show $\mu(B \cap C) = \mu(C)$. So let us assume not, and pick $\mathbf{p}, \mathbf{q}, D_{\mathbf{p}}, D_{\mathbf{q}}$ as in the previous proof. This time H is the *hemisphere* whose boundary bisects the arc $\mathbf{pq}$.

- Set $B^* := \text{foldup}_H(B)$, where the foldup is now defined w.r.t. a hemisphere in the obvious way. Everything works as before, showing $\mu(B^* \cap C) > \mu(B \cap C)$, *except* that we need to check that foldup_H cannot decrease the cut size: $\mu^2(\text{cut}(E^{(\vartheta)}, B^*)) \geq \mu^2(\text{cut}(E^{(\vartheta)}, B))$??

- Let us consider four unit vectors of the form $\mathbf{x}, \mathbf{y}, \sigma_H\mathbf{x}, \sigma_H\mathbf{y}$, where $\mathbf{x}, \mathbf{y} \in H$. Their angles satisfy $\angle\mathbf{xy} = \angle(\sigma_H\mathbf{x})(\sigma_H\mathbf{y}) \leq \angle\mathbf{x}(\sigma_H\mathbf{y}) = \angle(\sigma_H\mathbf{x})\mathbf{y}$.

- We introduce the *type* T of the ordered pair $(\mathbf{x}, \mathbf{y}) \in H \times H$, which specifies the following information: which of the four vectors $\mathbf{x}, \sigma_H\mathbf{x}, \mathbf{y}, \sigma_H\mathbf{y}$ lie in B, and which of the four angles considered above are at least ϑ. We can specify the type by a picture of the following kind:

This particular picture means that $\mathbf{x} \notin B$, while the other three points lie in B, and that $\angle\mathbf{x}(\sigma_H\mathbf{y}) = \angle(\sigma_H\mathbf{x})\mathbf{y}$ are at least ϑ, while the other two angles are smaller than ϑ.

- From the type of $(\mathbf{x}, \mathbf{y})$ one can determine which of the four pairs $(\mathbf{x}, \mathbf{y})$, $(\mathbf{x}, \sigma_H\mathbf{y})$, $(\sigma_H\mathbf{x}, \mathbf{y})$, $(\sigma_H\mathbf{x}, \sigma_H\mathbf{y})$ contribute to $\text{cut}(E^{(\vartheta)}, B)$ – those corresponding to bichromatic edges. So we have

$$\mu^2(\text{cut}(E^{(\vartheta)}, B)) = \sum_T k_T M_T,$$

where the sum is over all possible types, M_T is the μ^2-measure of all pairs of type T, and k_T is the number of bichromatic edges in the picture for type T.

- If we replace B by $B^* = \text{foldup}_H(B)$, each pair of type T becomes a pair of some type T^*, where T^* depends only on T. For example, for the type T depicted above, T^* is

- We have $M_T = M_{T^*}$, so it is enough to check that $k_{T^*} \geq k_T$ for all T, i.e., the new type is counted in the measure with at least as large a weight as the old type. There are just few easy cases to check.   □

8.3.40 Given this lemma, we still want to see that the best spherical cap is a hemisphere, i.e., $\mu^2(\text{cut}(E^{(\vartheta)}, H)) \geq \mu^2(\text{cut}(E^{(\vartheta)}, C))$. If we compute the cut measure in the Goemans–Williamson way, i.e., considering a random cap C of a given measure a and estimating the probability that it cuts a given $(\mathbf{x}, \mathbf{y}) \in E^{(\vartheta)}$, then this is a reasonably easy geometric argument. We leave it as an exercise.

Now, finally, we are done with Theorem 8.3.2! $\qquad\square$

8.4 The Approximation Ratio Is At Most α_{GW}

8.4.1 In Theorem 8.3.2, we constructed MAXCUT instances with large integrality gap, i.e., with the semidefinite optimum SDP as far from the true maximum cut Opt as it can ever be. But in such a case, the random hyperplane rounding finds an almost optimal cut.

Here we do a different thing: a graph where SDP = Opt, but where the random rounding finds only a cut of size about $\alpha_{\mathrm{GW}} \cdot$ Opt.

> **Theorem** (Karloff [Kar99]). *For every $\varepsilon > 0$ there exists a graph G with*
> $$\frac{\mathsf{Algo}}{\mathsf{Opt}} \leq \alpha_{\mathrm{GW}} + \varepsilon,$$
> *where Opt is the number of edges of a maximum cut and Algo is the expected size of the cut found by the random hyperplane rounding.*

- We will follow a somewhat simpler proof by Alon and Sudakov [AS00].
- Although the *expected* size of the hyperplane cut in this example is small, it turns out that the true maximum cut is also a hyperplane cut. It is not out of the question to modify the GW algorithm so that it finds the *best* hyperplane cut, instead of a random one. The theorem does not say anything about such an algorithm, but it is possible to modify the example so that *no* hyperplane is much better than a random one.

8.4.2 The construction of G has two parameters, an integer d and an even integer h. The ratio $\frac{h}{d}$ should approximate $(1 - \cos\vartheta_{\mathrm{GW}})/2 \approx 0.844579$. The ratio $\frac{\mathsf{Algo}}{\mathsf{Opt}}$ obtained for given d and h is $\frac{2}{\pi} \frac{1 - \cos\vartheta}{\vartheta}$, where $(1 - \cos\vartheta)/2 = \frac{h}{d}$.

8.4.3 G is the (binary) *Hamming graph* with vertex set

$$V := \{-1, 1\}^d$$

and edge set

$$E := \{\{a, b\} : a, b \in V, d_H(a, b) = h\}.$$

Here $d_H(a, b)$ is the *Hamming distance*, i.e., the number of coordinates where a and b differ.

8.4.4 With every vertex $a \in \{-1,1\}^d$ of G we associate the unit vector $\mathbf{v}_a := \frac{1}{\sqrt{d}} a \in \mathbb{R}^d$. Then $\mathbf{v}_a^T \mathbf{v}_b = 1 - \frac{2h}{d}$ for all edges $\{a, b\}$.

8.4.5 This obviously yields a *feasible* solution of the SDP relaxation. But this time we must exhibit an *optimal* solution, one that the algorithm could actually use for random rounding.

Lemma. *The just defined system of vectors* $(\mathbf{v}_a : a \in V)$ *is an optimal solution of the vector program* (GW); *i.e., it maximizes* $\sum_{\{a,b\} \in E} \frac{1 - \mathbf{v}_a^T \mathbf{v}_b}{2}$ *(assuming h even and $\frac{h}{d} \in [0.6, 0.9]$).*

8.4.6 Assuming this lemma, the theorem follows:

- Algo approaches $(\vartheta_{\mathrm{GW}}/\pi)|E|$ as $\frac{h}{d} \to (1 - \cos \vartheta_{\mathrm{GW}})/2$.
- Every cut by a coordinate hyperplane has $\frac{h}{d}|E|$ edges, and this is a lower bound for Opt. So Algo/Opt $\to \alpha_{\mathrm{GW}}$.
- Actually, SDP $= \frac{h}{d}|E|$ as well, and so SDP = Opt.

8.4.7 The lemma is proved by eigenvalue computation, using the following general fact.

Proposition. *Let $G = (V, E)$ be a graph on n vertices, let $A = A_G$ be its adjacency matrix and let $\lambda_{\min}$ be the smallest eigenvalue of A (most negative, not with a small absolute value; typically $\lambda_{\min} < 0$). Then*

$$\mathsf{SDP}(G) \leq \frac{1}{2}|E| + \frac{-\lambda_{\min} n}{4}.$$

8.4.8 Proof:

- Let $\mathbf{v}_1, \ldots, \mathbf{v}_n \in \mathbb{R}^n$ be an optimal SDP solution (unit vectors). We form the matrix with the $\mathbf{v}_i$ as columns, and let $\mathbf{r}_1, \ldots, \mathbf{r}_n \in \mathbb{R}^n$ be its *rows*.
- We have $\sum_{k=1}^n \|\mathbf{r}_k\|^2 = \sum_{i=1}^n \|\mathbf{v}_i\|^2 = n$.
- We compute

$$
\begin{aligned}
\mathsf{SDP} &= \frac{1}{2} \sum_{i,j=1}^n a_{ij} \frac{1 - \mathbf{v}_i^T \mathbf{v}_j}{2} = \frac{1}{2}|E| - \frac{1}{4} \sum_{i,j=1}^n \sum_{k=1}^n a_{ij} v_{ik} v_{jk} \\
&= \frac{1}{2}|E| - \frac{1}{4} \sum_{k=1}^n \mathbf{r}_k^T A \mathbf{r}_k.
\end{aligned}
$$

- For every $\mathbf{x} \in \mathbb{R}^n$, $\mathbf{x}^T A \mathbf{x} \geq \lambda_{\min} \|\mathbf{x}\|^2$. Equality holds iff $\mathbf{x}$ is an eigenvector of A belonging to $\lambda_{\min}$ (by the variational characterization of eigenvalues). This is a simple corollary of Theorem 4.8.1.
- So SDP $\leq \frac{1}{2}|E| - \frac{1}{4}\lambda_{\min} \sum_{k=1}^n \|\mathbf{r}_k\|^2 = \frac{1}{2}|E| - \frac{1}{4}\lambda_{\min} n$; equality holds iff all the $\mathbf{r}_k$ are eigenvectors belonging to $\lambda_{\min}$. □

Actually, the proposition can also be obtained from the SDP duality.

8.4.9 So what is $\lambda_{\min}$ of our (Hamming) graph? (For this method to show optimality of our SDP solution, the *rows* of B in the above proof must be eigenvectors of the adjacency matrix ... fortunately they are.)

Eigenvalues of Cayley Graphs

8.4.10 A reminder:

- Let Γ be an abelian ($=$ commutative) group.
- A *character* of Γ is a homomorphism $\chi\colon \Gamma \to \mathbb{C}^*$ (the group of all nonzero complex numbers with multiplication; really only $z \in \mathbb{C}$ with $|z| = 1$ are used).
- It is easily seen that all characters of Γ form a group $\hat{\Gamma}$ under pointwise multiplication. (This is important in Fourier analysis.)
- Now if Γ is an arbitrary (not necessarily abelian) group, and $S \subseteq \Gamma$ is a *symmetric set* (i.e., $s \in S$ implies $s^{-1} \in S$), the *Cayley graph* of Γ with respect to S is
$$(\Gamma, \{\{g, gs\} : g \in \Gamma, s \in S\}).$$
- Our Hamming graph is a Cayley graph: the group $\Gamma = (\{-1,1\}^d, \cdot)$ (coordinate wise multiplication), and
$$S = \{s \in \{-1,1\}^d : s \text{ has exactly } h \text{ components} -1\}.$$

8.4.11 Proposition. *Let G be the Cayley graph of Γ w.r.t. some S, where Γ is finite abelian, and let χ be a character of Γ. Then the vector $\boldsymbol{\chi} := (\chi(g) : g \in \Gamma)$ is an eigenvector of the adjacency matrix A of G, with eigenvalue $\lambda_\chi := \sum_{s \in S} \chi(s)$.*

8.4.12 Proof: $(A\boldsymbol{\chi})_g = \sum_{s \in S} \chi(gs) = \sum_{s \in S} \chi(s)\chi(g) = \lambda_\chi \chi(g).$ □

8.4.13 We claim that for our particular $\Gamma = (\{-1,1\}^d, \cdot)$, the characters are of the form χ_I with $\chi_I(g) = \prod_{i \in I} g_i$, where $I \subseteq \{1, 2, \ldots, d\}$. (This is a starting point of Fourier analysis on the Boolean hypercube.)

- Clearly, these χ_I are characters.
- In general, it is known that $|\hat{\Gamma}| = |\Gamma|$, and so these are *all* characters.

8.4.14 As we will see next, we can find all eigenvalues of the matrix A via Proposition 8.4.11.

- Indeed, it is easy to check that the characters χ_I as above are linearly independent as functions $\Gamma \to \mathbb{C}$.
- So we have 2^d linearly independent eigenvectors for a $2^d \times 2^d$ matrix – thus, we have *all* eigenvectors.

8.4.15 We compute the eigenvalue for χ_I, $|I| = k$:

$$\lambda_{\chi_I} = \sum_{s \text{ has } h \text{ components } -1} \chi_I(s) = \sum_{|J|=h} (-1)^{|I \cap J|}$$

$$= \sum_t (-1)^t \binom{k}{t} \binom{d-k}{h-t}$$

$$= K_h^d(k) \ldots \text{ the } binary \ Krawtchuk \ polynomial.$$

(This polynomial haunts the theory of error-correcting codes; see under *Delsarte's bound*.)

8.4.16 So $\lambda_{\min} = \min\left\{ K_h^d(k) : k = 0, 1, \ldots, d \right\}$.

8.4.17 Fact: the minimum occurs for $k = 1$ (for h *even* and such that $\frac{h}{d}$ is between 0.6 and 0.9, say).

- For not too large specific values of d and h, one can verify this by an exact calculation on a computer. For example, the exact numerical computation for $d = 1319$ and $h = 1114$ is quite feasible, and it shows the optimality of the Goemans–Williamson ratio to 6 decimal places.
- The proof for general d and j uses many estimates; it is a good mathematical craft, but lengthy and perhaps not so enlightening. So we omit it.

8.4.18 If we accept the just mentioned fact, we get $\lambda_{\min} = K_h^d(1)$. The eigenvectors belonging to $\lambda_{\min}$ are the characters χ_I with $|I| = 1$. Some thought reveals that these are, up to scaling, exactly the rows of the matrix made of the $\mathbf{v}_a$ (the SDP solution in Lemma 8.4.5).

8.4.19 As we saw in the proof of Proposition 8.4.7, for an SDP solution consisting of eigenvectors of the adjacency matrix A belonging to $\lambda_{\min}$, the inequality in the proposition holds with equality, and consequently, such an SDP solution is optimal. Lemma 8.4.5 follows, as well as Theorem 8.4.1. $\square$

8.5 The Unique Games Conjecture for Us Laymen, Part I

8.5.1 The Unique Games Conjecture is one of the most important open problems in theoretical computer science.

- *Proving it* would establish many amazingly precise results on hardness of approximation. For example, it would follow that no polynomial-time algorithm can approximate MAXCUT with ratio better than α_{GW}. We will state several other consequences later on.

- *Disproving it* would probably bring a powerful new algorithmic technique, which might be useful in a number of other problems.
- But perhaps proof or disproof is very hard, the conjecture might remain open for a long time, and we will have to live with it, as we do with $P \neq NP$.

8.5.2 Here we state the Unique Games Conjecture in a simple and concrete form. But much more material is needed to explain the origins of the conjecture and its connections to inapproximability – at least another one-semester course.

8.5.3 We introduce the algorithmic problem **MAX-2-LIN(mod q)**, where q is a fixed prime. The *input* is a system of m linear equations modulo q, with unknowns $x_1, \ldots, x_n \in \{0, 1, \ldots, q-1\}$, where each of the equations has a very simple form $x_i - x_j = c$. For example,

$$x_3 - x_{11} \equiv 87 \pmod{97}$$
$$x_7 - x_{22} \equiv 3 \pmod{97}$$
$$\vdots \quad \vdots \quad \vdots$$
$$x_7 - x_{19} \equiv 56 \pmod{97}.$$

Task: Find an assignment of the x_i that satisfies the maximum possible number of the given equations.

8.5.4 A $(1-\delta, \varepsilon)$-*gap version* of MAX-2-LIN(mod q): If there is an assignment satisfying at least a $(1 - \delta)$-fraction of the equations, the output should be YES; if no assignment satisfies more than an ε-fraction of the equations, the output should be NO; in the remaining cases, the output can be either YES or NO.

> *Unique Games Conjecture*, concrete form.
>
> *For every $\varepsilon > 0$ there exists a prime q such that the $(1 - \varepsilon, \varepsilon)$-gap version of* MAX-2-LIN(mod q) *admits no polynomial-time algorithm unless* $P = NP$.

- By the way, where are the games, and why unique? Better do not ask (this refers to a different, essentially equivalent, and more complicated setting...).

8.5.5 Unlike $P \neq NP$, the UGC looks somewhat arbitrary; there do not seem to be any compelling reasons why it should (or should not) be true.

- Some recent work, most notably a paper by Arora et al. [ABS10], perhaps points more towards a possible refutation of the UGC. Namely, this paper gives an approximation algorithm for a (generalization of) MAX-2-LIN(mod q) that runs in time less than exponential in m, the number of equations.

- Concretely, for some (large) constant C, they give an ingenious algorithm that, given an instance of Max-2-Lin(mod q) that has an assignment satisfying at least $(1 - \varepsilon)m$ of the equations (for some $\varepsilon > 0$), outputs an assignment satisfying at least $(1 - \varepsilon^C)m$ of the equations. The running time is bounded by $\exp(O(qm^{\varepsilon^C}))$.

8.5.6 But, the UGC has gained a central status because of the numerous consequences. In particular, as we mentioned, it implies that α_{GW} is the optimal approximation ratio for MaxCut.

8.5.7 The proof of this implication by Khot et al. [KKMO07] relies on techniques around the PCP theorem and uses advanced Fourier analysis on the Boolean cube.

- The technical core of this result, as well as of a number of others, is a theorem in discrete Fourier analysis called *"majority is stablest."*
- In geometric terms, the "majority is stablest" theorem says essentially that in the Hamming graph example discussed in Sect. 8.4, every near-optimal cut is close to a cut by a coordinate hyperplane, while all other cuts are about as bad as a random hyperplane cut. See O'Donnell's lecture notes, e.g., http://www.cs.cmu.edu/~odonnell/boolean-analysis/.

8.5.8 In 2010, Guruswami et al. [GRSW10] obtained tight inapproximability results for two geometric computational problems (which we will not describe here) assuming only $P \neq NP$. Previously, the same inapproximability bounds were derived assuming the UGC as well. Perhaps there is hope to bypass the UGC in a similar way for some of the other optimal inapproximability results.

Exercises

8.1 Let G be a bipartite graph. What is the optimum value of the vector program (GW), and what does the corresponding vector representation look like?

8.2 Prove that the five vectors arranged in a regular pentagon, as shown in the picture in 8.3.3, constitute an *optimal* solution of the vector program (GW) for the 5-cycle C_5.

8.3 Suppose that for some graph G, the Goemans–Williamson vector program (GW) has an optimal solution whose vectors are all contained in the set $\{-d^{-1/2}, d^{-1/2}\}^d \subset \mathbb{R}^d$ (for some $d \leq n$). Prove that then the integrality gap equals 1; that is, there exists a cut in G whose number of edges equals the optimum of (GW).

8.4 Consider an extension of (GW) with the following "triangle constraints" added:

$$\|\mathbf{v}_i - \mathbf{v}_j\|^2 + \|\mathbf{v}_j - \mathbf{v}_k\|^2 \geq \|\mathbf{v}_i - \mathbf{v}_k\|^2 \quad \text{for all } i, j, k$$
$$\|\mathbf{v}_i + \mathbf{v}_j\|^2 + \|\mathbf{v}_j + \mathbf{v}_k\|^2 \geq \|\mathbf{v}_i - \mathbf{v}_k\|^2 \quad \text{for all } i, j, k.$$

(i)   Check that the first constraint is equivalent to

$$\mathbf{v}_i^T \mathbf{v}_k - \mathbf{v}_i^T \mathbf{v}_j - \mathbf{v}_j^T \mathbf{v}_k \geq -1.$$

(ii) Prove that for every cycle C_n, the integrality gap of this extended vector program is 1, i.e., the optimum value is the true size of the maximum cut.

(iii) What is the expected size of a cut obtained by the random hyperplane method (the same as in the Goemans–Williamson algorithm) for an optimal solution of this extended vector program for C_7? Compare it to the ratio α_{GW}. (Hint: Find an optimal solution with vectors $\mathbf{v}_i$ in the set $\{-d^{-1/2}, d^{-1/2}\}^d$.)

8.5 Prove Lemma 8.3.35(i) rigorously (elementary measure theory may be needed.)

8.6 Prove Lemma 8.3.35(iii) following the hint in the proof sketch.

8.7 Prove the last claim in 8.3.40, i.e., $\mu^2(\text{cut}(E^{(\vartheta)}, H)) \geq \mu^2(\text{cut}(E^{(\vartheta)}, C))$, where C is an arbitrary cap and H is a hemisphere.

Chapter 9
Coloring 3-Chromatic Graphs

9.1 The 3-Coloring Challenge

9.1.1 We recall

- The *chromatic number* $\chi(G)$, the smallest number of colors needed to color the vertices of a graph G so that no two neighboring vertices receive the same color.
- The *independence number* $\alpha(G)$, the size of the largest independent set in G, where an independent set is one with no two of its vertices connected by an edge.

9.1.2 Both of these graph parameters are computationally difficult: they are NP-hard to approximate with factor $n^{1-\varepsilon}$, for every fixed $\varepsilon > 0$.

- Feige and Kilian [FK98] and Håstad [Hås99] obtained this kind of result with a randomized reduction, which was then derandomized by Zuckerman [Zuc06]; also see Khot and Ponnuswami [KP06] for a still slightly stronger inapproximability result.

9.1.3 The strongest known positive results, i.e., efficient coloring algorithms, are based on SDP.

9.1.4 We will consider the following particular *challenge* in graph coloring: Assuming that $\chi(G) = 3$, color G in polynomial time with as few colors as possible. (There are obvious generalizations for $k > 3$ fixed but we stick to the case $k = 3$.)

9.1.5 Known hardness results for this case:

- If $P \neq NP$, then there is no polynomial-time algorithm guaranteed to color every 3-colorable graph with at most 4 colors (Khanna et al. [KLS00]).
- Khot's "2-to-1 conjecture," similar to the Unique Games Conjecture, implies that there is no polynomial-time algorithm guaranteed to color

B. Gärtner and J. Matoušek, *Approximation Algorithms and Semidefinite Programming*, DOI 10.1007/978-3-642-22015-9_9,
© Springer-Verlag Berlin Heidelberg 2012

every 4-colorable graphs with one million (or any other fixed number) of colors; see Dinur et al. [DMR06].

9.1.6 Obviously, every n-vertex graph can be colored with n colors. Here is a simple way of using significantly fewer colors, namely, $O(\sqrt{n})$ (*Wigderson's trick*):

- If G is 3-colorable, then the neighborhood of a vertex is bipartite and thus it can be 2-colored (fast).
- So, given a parameter Δ, by repeatedly coloring maximum-degree vertices and their neighbors, we can use $O(n/\Delta)$ colors and reduce the maximum degree below Δ.
- A graph with maximum degree Δ can easily be colored by $\Delta + 1$ colors (greedy algorithm – give each vertex a color not used by its neighbors).
- Combining these two methods and setting $\Delta := \sqrt{n}$, we can color every 3-colorable G with $O(\sqrt{n})$ colors.

9.1.7 Other tricks, ingenious but still elementary, can get this down to $\tilde{O}(n^{0.375})$ (Blum [Blu94]). This was the record for about 10 years.

- We recall that $\tilde{O}(.)$ is a convenient notation for ignoring logarithmic factors; $f(n) = \tilde{O}(g(n))$ means $f(n) \leq g(n)(\log n)^{O(1)}$.

9.1.8 This barrier was finally broken using SDP by Karger et al. [KMS98]: $\tilde{O}(n^{0.25})$ colors suffice – this is what we are now going to present.

9.1.9 Some improvement of the $\tilde{O}(n^{0.25})$ bound can be achieved by employing the Blum tricks.

9.1.10 A conceptual improvement of the Karger–Motwani–Sudan algorithm (a "second generation," global analysis of the rounding step) was given by Arora et al. [ACC06], achieving $\tilde{O}(n^{0.2111})$ colors.

9.1.11 A further slight improvement to $\tilde{O}(n^{0.2072})$ was obtained by Chlamtac [Chl07], who used an SDP relaxation at a higher level of the *Lasserre hierarchy* (which we will briefly mention in 12.3.9). Here we will not discuss these more advanced results and cover only the Karger–Motwani–Sudan algorithm, which constitutes a basis for all further developments.

9.2 From a Vector Coloring to a Proper Coloring

9.2.1 A *(non-strict) vector k-coloring* of a graph G is an assignment of unit vectors $\mathbf{v}_i$ to vertices such that for every two adjacent vertices i, j we have

$$\mathbf{v}_i^T \mathbf{v}_j \leq -\frac{1}{k-1}.$$

- The smallest k can be found using SDP (k need not be an integer); see Chap. 3.
- Every k-colorable graph has a vector k-coloring. So for 3-colorable G, we may assume a vector 3-coloring.

  ○ Earlier we have seen *strict vector coloring*, with $\mathbf{v}_i^T \mathbf{v}_j = -\frac{1}{k-1}$. There the smallest k equals $\vartheta(\overline{G})$, the Lovász theta function of the complement.
  ○ It not clear whether the vector chromatic number and the strict vector chromatic number may ever differ – at least no such example seems to be known.

9.2.2 The basis of the KMS coloring algorithm is the following result.

Theorem (The Karger–Motwani–Sudan rounding [KMS98]). *There is a polynomial-time randomized algorithm which, given a graph G on n vertices of maximum degree at most Δ and a vector 3-coloring of G, finds an independent set in G whose expected number of vertices is bounded by $\tilde{\Omega}(\Delta^{-1/3} n)$.*

9.2.3 How can this be used for a *coloring*? Having found an independent set, we assign it a new color, and color the rest of the graph recursively. So any graph as in the theorem can be colored with $\tilde{O}(\Delta^{1/3})$ colors.

9.2.4 Combining Theorem 9.2.2 with Wigderson's trick (see 9.1.6) and setting $\Delta := n^{3/4}$ (we leave the details as an exercise), we obtain:

Corollary. *A 3-colorable graph can be colored with $\tilde{O}(n^{0.25})$ colors in randomized polynomial time.*

(Adding Blum's tricks even gives $\tilde{O}(n^{0.2143})$ colors.)

9.3 Properties of the Normal Distribution

9.3.1 Now we make a detour. First we recall the *standard normal* or *Gaussian distribution*, denoted by $N(0,1)$.

- The *density* of the standard normal distribution is $\varphi(x) := \frac{1}{\sqrt{2\pi}} e^{-x^2/2}$, the famous *bell curve*.

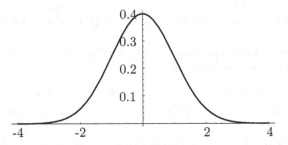

That is, if $Z \sim N(0,1)$ (read "Z is a random variable with the standard normal distribution"), then for every $x \in \mathbb{R}$, $\mathrm{Prob}[Z \in [x, x+h]]$ is approximately $h \cdot \varphi(x)$ for h small.

◦ More precisely,

$$\lim_{h \to 0} \frac{1}{h} \mathrm{Prob}[Z \in [x, x+h]] = \varphi(x).$$

• We will use the function $N(t) := \mathrm{Prob}[Z \geq t] = \int_t^\infty \varphi(x)\,\mathrm{d}x$, where $Z \sim N(0,1)$.

9.3.2 We will need that, for t large, $N(2t)$ is approximately $N(t)^4$.

• This is how the density function behaves: $\varphi(2t) = (2\pi)^{3/2}\varphi(t)^4$.
• Intuitively, for large t, the behavior of $N(t)$ is similar to that of $\varphi(t)$, and that "explains" why $N(2t) \approx N(t)^4$. But, in order to derive a rigorous result of this kind, we need good estimates of $N(t)$.
• It is known that $N(t)$ cannot be expressed using elementary functions. In other words, the integral of the density function $\varphi(x)$ cannot be computed by the usual tricks from calculus; it is a "new" function. The main trick for estimating $N(t)$ is finding another function, which is "close" to $\varphi(x)$ but can be integrated. Namely, we use the following two functions:

◦ $(1 + \frac{1}{x^2})\varphi(x)$ as an upper bound for $\varphi(x)$.

◦ $(1 - \frac{1}{3x^4})\varphi(x)$ as a lower bound.

Both of them can be integrated by parts, which we leave to the reader.

This yields the following useful estimates.

Lemma. *For all $t \geq 0$, we have*

$$\left(\frac{1}{t} - \frac{1}{t^3}\right)\frac{1}{\sqrt{2\pi}}e^{-t^2/2} \leq N(t) \leq \frac{1}{t}\frac{1}{\sqrt{2\pi}}e^{-t^2/2}.$$

9.3.3 We will also need the n-*dimensional standard normal* (or *Gaussian*) *distribution*.

- This is the distribution of a random vector $\gamma = (\gamma_1, \gamma_2, \ldots, \gamma_n) \in \mathbb{R}^n$, whose components $\gamma_1, \ldots, \gamma_n$ are independent $N(0,1)$ random variables.
- The distribution of γ is *spherically symmetric*: its density function is

$$(2\pi)^{-n/2} \prod_{i=1}^{n} e^{-x_i^2/2} = (2\pi)^{-n/2} e^{-\|\mathbf{x}\|^2/2}.$$

- Thus, if $\mathbf{u}$ is a unit vector, then $\gamma^T \mathbf{u} \sim N(0,1)$ (since this holds, by definition, for $\mathbf{u} = \mathbf{e}_1$).
- It is hard to overstate the importance of the n-dimensional Gaussian distribution in probability theory and statistics, but also in geometry and algorithms. In this book, which is certainly not intentionally focused on probabilistic methods, we will meet the Gaussian distribution in several rather different algorithms.

9.4 The KMS Rounding Algorithm

9.4.1 Let G be a graph with maximum degree Δ, and let $\mathbf{v}_1, \ldots, \mathbf{v}_n \in \mathbb{R}^n$ be a vector 3-coloring (unit vectors, $\mathbf{v}_i^T \mathbf{v}_j \leq \frac{1}{2}$ for all edges). We want to find a large independent set I in G. The Karger–Motwani–Sudan algorithm proceeds as follows.

- Pick $\gamma \in \mathbb{R}^n$ random Gaussian, and let $I_0 := \{i : \gamma^T \mathbf{v}_i \geq t\}$, t a suitable threshold parameter.
- Let $I \subseteq I_0$ be the set of all isolated vertices in I_0 (having no neighbor in I_0). This is the desired independent set.

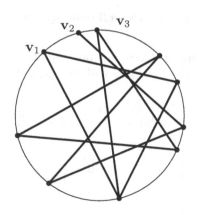
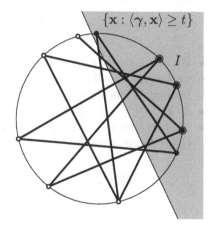

9.4.2 Intuition: If $\{i, j\} \in E$, then $\mathbf{v}_i$ and $\mathbf{v}_j$ are far away, and thus unlikely to fall in the same cap.

9.4.3 Analysis:

- $\mathbf{E}\left[|I|\right] = \mathbf{E}\left[|I_0|\right] - \mathbf{E}\left[|I_0 \setminus I|\right].$
- $\mathbf{E}\left[|I_0|\right] = \sum_{i=1}^{n} \mathrm{Prob}[i \in I_0] = \sum_{i=1}^{n} \mathrm{Prob}\left[\boldsymbol{\gamma}^T \mathbf{v}_i \geq t\right] = nN(t).$
- We calculate

$$
\begin{aligned}
\mathbf{E}\left[|I_0 \setminus I|\right] &= \sum_{i=1}^{n} \mathrm{Prob}[i \in I_0 \text{ and } j \in I_0 \text{ for some edge } \{i,j\}] \\
&\leq \sum_{i=1}^{n} \sum_{\{i,j\} \in E} \mathrm{Prob}[i,j \in I_0] \quad \text{(by the union bound)}.
\end{aligned}
$$

- We have $\mathrm{Prob}[i,j \in I_0] = \mathrm{Prob}\left[\boldsymbol{\gamma}^T \mathbf{v}_i \geq t \text{ and } \boldsymbol{\gamma}^T \mathbf{v}_j \geq t\right]$; geometrically, we ask for the probability of $\boldsymbol{\gamma}$ falling in the dark gray wedge:

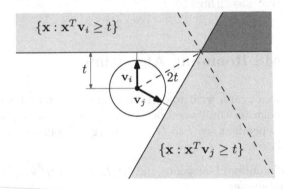

- For $\mathbf{v}_i$ fixed, the wedge grows larger as the angle $\angle \mathbf{v}_i \mathbf{v}_j$ decreases. But the vector coloring condition tells us that $\mathbf{v}_i^T \mathbf{v}_j \leq -\frac{1}{2}$, so $\angle \mathbf{v}_i \mathbf{v}_j$ is at least $120°$. We may assume that it is exactly that.

  ○ Alternatively, we could have started with a *strict* vector 3-coloring; then $\mathbf{v}_i^T \mathbf{v}_j = -\frac{1}{2}$.

- We could calculate the probability of $\boldsymbol{\gamma}$ lying in the wedge precisely, but we estimate it: The wedge is contained in the halfspace with the dotted boundary, whose distance from the origin is $2t$.
- So $\mathrm{Prob}[i,j \in I_0] \leq N(2t)$, and $\mathbf{E}\left[|I|\right] \geq n(N(t) - \Delta N(2t))$.
- Set t so that the last expression is large; a good choice is $t := (\frac{2}{3} \ln \Delta)^{1/2}$. Using Lemma 9.3.2 gives

$$
\begin{aligned}
N(t) - \Delta N(2t) &\geq \frac{1}{\sqrt{2\pi}} \left(\left(\frac{1}{t} - \frac{1}{t^3} \right) e^{-t^2/2} - \frac{\Delta}{2t} e^{-4t^2/2} \right) \\
&= \Omega\left(\Delta^{-1/3}/\sqrt{\ln \Delta} \right)
\end{aligned}
$$

(we may assume that Δ is larger than a suitable constant, and thus $t \geq 2$, say).

- So the expected size of the independent set I is indeed $\tilde{\Omega}(\Delta^{-1/3}n)$, and Theorem 9.2.2 is proved. $\square$

9.5 Difficult Graphs

9.5.1 We will construct a graph G with vector chromatic number at most 3 but chromatic number large. Such G shows that it is impossible to guarantee coloring with a very small number of colors based only on vector chromatic number 3.

> **Proposition.** *There exists a constant $\delta > 0$ such that for infinitely many values of n, one can construct an n-vertex graph with vector chromatic number at most 3 and with chromatic number at least n^δ.*

- The best δ is obtained by a construction similar to the one for the integrality gap of the Goemans–Williamson algorithm. But that is quite complicated.
- Here we explain a weaker result with a very neat proof (two proofs, actually).

9.5.2 Construction: The graph G has all s-element sets $A \subseteq \{1, 2, \ldots, d\}$ as vertices, and edges correspond to pairs of sets A, B with $|A \cap B| \leq t$; the parameters s, t need to be chosen carefully. Setting $d = 8t$ and $s = 4t$ will work. Thus $n = |V(G)| = \binom{d}{s}$.

9.5.3 Let us see how the vector chromatic number works; this is simple.

- If $A \in V(G)$ is an s-element set, we assign it the unit vector $\mathbf{v}_A \in \mathbb{R}^d$ (normalized signed characteristic vector): $(\mathbf{v}_A)_i := d^{-1/2}$ if $i \in A$ and $(\mathbf{v}_A)_i := -d^{-1/2}$ if $i \notin A$.
- $\{A, B\} \in E(G)$ means $|A \cap B| \leq t$; then $\mathbf{v}_A^T \mathbf{v}_B \leq \frac{1}{d}(d - 4s + 4t)$.
- For $d = 8t$ and $s = 4t$ as above, we get $\mathbf{v}_A^T \mathbf{v}_B \leq -\frac{1}{2}$, i.e., vector chromatic number at most 3.

9.5.4 It remains to bound the chromatic number $\chi(G)$ from below.

- We will apply the "usual" lower bound $\chi(G) \geq n/\alpha(G)$. (This holds because a coloring of G with k colors means a covering of the vertex set with k independent sets.)
- Here $\alpha(G)$ is the maximum independent set size; in our case, the maximum number of s-element subsets of $\{1, 2, \ldots, d\}$ such that every two intersect in at least $t + 1$ elements.
- For bounding $\alpha(G)$ from above, will need a "forbidden intersections" theorem. There are many beautiful "forbidden intersections" theorems about set systems, which bound the size of the maximum independent set

in graphs of this kind (Ray-Chaudhuri–Wilson, Frankl–Wilson, Frankl–Rödl ...). We will rely on the following theorem.

9.5.5 Theorem. Let $\mathcal{F}$ be a system of s-element subsets of $\{1, 2, \ldots, d\}$ such that every two distinct $A, B \in \mathcal{F}$ satisfy $|A \cap B| \geq t + 1$. Then

$$|\mathcal{F}| \leq \binom{d}{0} + \binom{d}{1} + \cdots + \binom{d}{s - t - 1}.$$

- This resembles the famous *Erdős–Ko–Rado theorem*.
- The well-known basic version of that theorem asserts that if $\mathcal{F}$ is a system of k-element subsets of an n-element set, with $A \cap B \neq \emptyset$ for every $A, B \in \mathcal{F}$, and $n \geq 2k$, then $|\mathcal{F}| \leq \binom{n-1}{k-1}$.
- Erdős, Ko, and Rado also proved that if $|A \cap B| \geq t+1$ for every $A, B \in \mathcal{F}$, then $|\mathcal{F}| \leq \binom{n-t-1}{k-t-1}$ *provided* that $n \geq n_0(k, t)$, where $n_0(s, t)$ has to be sufficiently large – the smallest value is known to be $(k - t)t$. So in our setting EKR is *not applicable*!
- But we can get the slightly worse bound in the theorem.

9.5.6 Let us first see how Theorem 9.5.5 implies Proposition 9.5.1. By the above, it suffices to bound $\alpha(G)$ from above.

- We employ the following generally useful approximation for the binomial coefficient: for $\beta \in (0, 1)$ and n large, we have

$$\binom{n}{\beta n} \approx 2^{nH(\beta)},$$

where $H(x) := -x \log_2 x - (1 - x) \log_2(1 - x)$ is the *entropy function*. More precisely, for every $\beta \in (0, 1)$ fixed, we have $\lim_{n \to \infty} \frac{1}{n} \log_2 \binom{n}{\lfloor \beta n \rfloor} = H(\beta)$. (This is not so hard to prove but we skip that.)
- Thus, for t large, the bound in Theorem 9.5.5 is about $2^{dH(3/8)} \approx 2^{0.954d}$. That is an upper bound on $\alpha(G)$.
- Our $n = \binom{d}{s} \approx 2^d$, and $\chi(G) \geq n/\alpha(G) \geq 2^{0.045d}$. Then we obtain $\delta \approx 0.045$, and the proposition is proved. □

9.5.7 It remains to prove Theorem 9.5.5. The *first proof* is based on a linear algebra trick – the *polynomial method* according to Alon et al. [ABS91].

- To each $A \in \mathcal{F}$ we assign two things:

  ○ A vector $\mathbf{c}_A \in \{0, 1\}^d$: the characteristic vector of A (a 0/1 vector this time! sorry), with the i-th component 1 if $i \in A$ and 0 otherwise.

○ A function $f_A: \{0,1\}^d \to \mathbb{R}$, given by

$$f_A(\mathbf{x}) = \prod_{j=t+1}^{s-1} \left(\left(\sum_{i \in A} x_i \right) - j \right).$$

- We see: $f_A(\mathbf{c}_A) \neq 0$, $f_A(\mathbf{c}_B) = 0$ for $B \neq A$.
- This implies: The f_A are *linearly independent* (as elements of the vector space of all maps $\{0,1\}^d \to \mathbb{R}$).

  ○ Indeed, we consider a linear combination $\sum_{A \in \mathcal{F}} \alpha_A f_A$ equal to the zero function, and substitute $\mathbf{c}_B$ into it. This yields $\alpha_B = 0$, and since B was arbitrary, all the α_A are zero.

- Next, we want to show that all the f_A lie in a vector space generated by a small number of functions. Suitable functions are given by monomials, of the form $x_1^{i_1} x_2^{i_2} \cdots x_d^{i_d}$.
- Each f_A is a polynomial in $x_1, \ldots, x_d$ of degree at most $s - t - 1$, i.e., a linear combination of monomials.
- Since we are dealing with functions defined on $\{0,1\}^d$, x_i^2 defines the same function as x_i. So we can consider only *squarefree* (or *multilinear*) monomials, in which every variable has power 0 or 1.
- It is easily counted that there are $\binom{d}{0} + \binom{d}{1} + \cdots + \binom{d}{s-t-1}$ squarefree monomials of degree $\leq s - t - 1$ in d variables. They generate a vector space W containing all f_A. Since we know that the f_A are linearly independent, their number is no more than $\dim W$. The theorem is proved. □

9.5.8 The *second proof* is based on the (generally useful) Vapnik–Chervonenkis–Sauer–Shelah lemma, which we now state.

- Let $\mathcal{M}$ be a set system on a set X. A finite set $S \subseteq X$ (not necessarily belonging to $\mathcal{M}$) is called *shattered* by $\mathcal{M}$ if for every $T \subseteq S$ there exists $M \in \mathcal{M}$ with $S \cap M = T$.
- The *Vapnik–Chervonenkis–Sauer–Shelah lemma*: If X and $\mathcal{M}$ are as above, $|X| = n$, and no set $S \subseteq X$ with *more than* k elements is shattered by $\mathcal{M}$, then $\mathcal{M}$ contains at most $\binom{n}{0} + \binom{n}{1} + \cdots + \binom{n}{k}$ sets.
- This lemma is not hard to prove by double induction (on k and on n).
- In the setting of Theorem 9.5.5, we check that no subset S of $\{1, 2, \ldots, d\}$ of size $s - t$ is shattered by $\mathcal{F}$. Indeed, if such an S were shattered, there is some $A \in \mathcal{F}$ with $S \cap A = \emptyset$, and there is also some $B \in \mathcal{F}$ with $S \cap B = S$. Then, however, $|A \cap B| \leq |B| - |S| = s - (s - t) < t + 1$, which contradicts the assumption on $\mathcal{F}$.
- Thus, Theorem 9.5.5 follows from the Vapnik–Chervonenkis–Sauer–Shelah lemma. □

Exercises

9.1 Given a graph G with n vertices and maximum degree Δ, we want to find a large independent set. Analyze the following randomized strategy.

First, we put each vertex v into a set I_0 independently with probability p (for a suitable parameter p). Next, we remove each adjacent pair from I_0, obtaining the independent set $I := \{v \in I_0 : v$ has no neighbors in $I_0\}$. What is its expected size? Which value of p maximizes this expectation?

How does the result change if we remove from I_0 only one vertex from each edge?

9.2 Suppose that an algorithm is given that, for every n-vertex 3-colorable graph with maximum degree Δ, finds an independent set of size at least $cn/\Delta^{1/3}$, where $c > 0$ is a constant. Show that using this algorithm, we can color every n-vertex 3-colorable graph with $\tilde{O}(n^{1/4})$ colors.

9.3 Prove Lemma 9.3.2, following the hints in the text.

9.4 Use the polynomial method as in 9.5.7 to prove the *Frankl–Wilson inequality*:

Let p be a prime, and let d and s be integers with $d > s \geq p$. Let $\mathcal{F}$ be a system of s-element subsets of $\{1, \ldots, d\}$ such that for every two distinct $A, B \in \mathcal{F}$, we have $|A \cap B| \not\equiv s \pmod{p}$. Then

$$|\mathcal{F}| \leq \sum_{i=0}^{p-1} \binom{d}{i}.$$

9.5 Prove the Vapnik–Chervonenkis–Sauer–Shelah lemma by double induction, on k and on n.

9.6 (a) Prove the inequality $\binom{n}{k} \leq 2^{nH(k/n)}$ for all $n \geq k \geq 1$, where $H(\cdot)$ is the entropy function (a true inequality, no asymptotics!).

(b) Prove the asymptotic formula in the text, namely,

$$\lim_{n \to \infty} \frac{\log_2 \binom{n}{\lfloor \alpha n \rfloor}}{n} = H(\alpha), \quad \alpha \in (0, 1).$$

This can be done using with the Stirling formula for $n!$; you may also want to find a proof avoiding it.

Chapter 10
Maximizing a Quadratic Form on a Graph

10.1 Four Problems

MaxCutGain

10.1.1 This is a reconsideration of the objective function in MaxCut.

- A cut with $\frac{1}{2}|E|$ edges may be regarded as "trivial" (this is the expected size if we partition the vertex set randomly); *real* designers of algorithms can do better. So we can perhaps better measure the quality of a MaxCut algorithm by the "gain"; MaxCut$-\frac{1}{2}|E|$.
- Solving MaxCutGain *exactly* is equivalent to solving MaxCut, but *approximation* is a very different story (e.g., if the maximum cut has $\frac{1}{2}|E| + |E|^{0.9}$ edges and an algorithm finds a cut with $\frac{1}{2}|E| + 10$ edges).

10.1.2 An integer quadratic program formulation of MaxCutGain:

$$\max\left\{ \sum_{\{i,j\}\in E} \frac{-x_i x_j}{2} : x_1,\ldots,x_n \in \{\pm 1\}\right\}$$

(cut edges contribute $+\frac{1}{2}$, non-cut edges $-\frac{1}{2}$).

Ground State in the Ising Model

10.1.3 The *Ising model*.

- V is a set of atoms (in a crystal)
- $E \subseteq \binom{V}{2}$ are the interacting pairs of atoms ("adjacent")

B. Gärtner and J. Matoušek, *Approximation Algorithms and Semidefinite Programming*, DOI 10.1007/978-3-642-22015-9_10,
© Springer-Verlag Berlin Heidelberg 2012

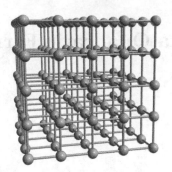

- *State* of the model: each atom has a *spin* $x_i \in \{+1, -1\}$
- *Energy* of the state $= -\sum_{\{i,j\} \in E} J_{ij} x_i x_j$, where J_{ij} is an *interaction constant* for the pair $\{i, j\}$. (For example, for modeling *ferromagnetism*, the J_{ij} are positive.)

10.1.4 This is an important, albeit very simplified, model in physics. It is much studied and not fully understood. It exhibits *phase transition* (in the limit for $|V| \to \infty$) and other remarkable phenomena.

10.1.5 One of the basic problems related to the Ising model is finding a *ground state*: i.e., we want to set the spins so that the energy is minimum.

Correlation Clustering

10.1.6 We are given a set $V = \{1, 2, \ldots, n\}$ of objects and *judgments* about their similarity, i.e., a function

$$\binom{V}{2} \to \{\text{similar}, \text{dissimilar}, \text{do not know}\}.$$

10.1.7 Wanted: a partition of V into *clusters* so that the *correlation* is maximized, where correlation = the number of "right" judgments minus the number of "wrong" ones. Here a right judgment is a *similar* pair in the same cluster or a *dissimilar* pair divided to different clusters. Wrong judgments are the opposite, and the "do not know" judgments are ignored.

10.1.8 It is not difficult to show that, up to an approximation factor of 3, it suffices to look at 2-clusterings and the (single) n-clustering (Exercise 10.1).

10.1.9 For a 2-clustering, maximizing the correlation can be expressed by the integer quadratic program

$$\max\left\{ \sum_{\{i,j\} \text{ similar}} x_i x_j - \sum_{\{i,j\} \text{ dissimilar}} x_i x_j : x_1, \ldots, x_n \in \{\pm 1\} \right\}.$$

CUTNORM

10.1.10 For an $m \times n$ matrix A, we define the *cut norm* $\|A\|_{\mathrm{cut}}$ as

$$\max\left\{ \Big| \sum_{i \in I, j \in J} a_{ij} \Big| : I \subseteq \{1, 2, \ldots, m\}, J \subseteq \{1, 2, \ldots, n\} \right\}.$$

- The cut norm is a key concept in the work of Frieze and Kannan [FK99] on approximation algorithms for dense graphs (e.g., MAXCUT for a graph with at least δn^2 edges, $\delta > 0$ fixed), with approximation factor arbitrarily close to 1 (i.e., $1 - \varepsilon$, where the running time depends exponentially on ε^{-1}).
- Using an algorithm for approximating the cut norm of a matrix, they produce a decomposition of a given matrix $A = C^{(1)} + C^{(2)} + \cdots + C^{(s)} + W$, where each $C^{(k)}$ is a "cut" matrix (entries t_k on some $I \times J$ and zeros elsewhere), and W is a small error term. Such a decomposition is computed "greedily," by approximating the cut norm of the current matrix at each step and subtracting an appropriate cut matrix $C^{(k)}$.
- Once we have such a decomposition, we can compute things for the $C^{(k)}$ and ignore W. This leads to simple approximation algorithms for several problems.

10.1.11 A formulation of CUTNORM as an integer quadratic program (trick): Given the matrix A, we first make a new matrix B by adding a new row $m+1$ and a new column $n+1$, where $b_{m+1,j} := -\sum_{k=1}^{m} a_{kj}$ (negative column sums), $b_{i,n+1} := -\sum_{k=1}^{n} a_{ik}$ (negative row sums), and $b_{m+1,n+1} := \sum_{i,j} a_{ij}$. Then we have

$$\|A\|_{\mathrm{cut}} = \|B\|_{\mathrm{cut}}$$

(exercise), and

$$\|B\|_{\mathrm{cut}} = \frac{1}{4} \max\left\{ \sum_{i=1}^{m+1} \sum_{j=1}^{n+1} b_{ij} x_i y_j : x_1, \ldots, x_{m+1}, y_1, \ldots, y_{n+1} \in \{\pm 1\} \right\}$$

for every matrix B with zero row sums and zero column sums (another exercise).

10.2 Quadratic Forms on Graphs

10.2.1 We motivated the following problem:

MAXQP[G]: maximizing a quadratic form on a graph
$$G = (V, E)$$

$$\max\left\{ \sum_{\{i,j\} \in E} a_{ij} x_i x_j : x_1, \ldots, x_n \in \{\pm 1\} \right\},$$

where a_{ij} are real weights on edges, generally both positive and negative.

10.2.2 This includes all of the four problems above.

- For CUTNORM, the graph G is *complete bipartite*. For the Ising model, the graph is usually sparse (e.g., 3-dimensional grid).

10.2.3 We want to write down a semidefinite relaxation of MAXQP[G].

- We assume that $G = (\{1, 2, \ldots, n\}, E)$ has *no loops* (that is, $a_{ii} = 0$). So we maximize a quadratic form *with no square terms*. Then the form is a *linear* function of each x_i. Consequently, we can as well consider $x_1, \ldots, x_n \in [-1, 1]$.
- Moreover, if someone gives us a fractional solution, with some $x_i \in (-1, 1)$, we can easily find a solution with all $x_i \in \{-1, 1\}$ at least as good.
- The following SDP relaxation of MAXQP[G] is essentially the same as that for MAXCUT:

SDP relaxation of MAXQP[G]

$$S_{\max} := \max\left\{ \sum_{\{i,j\} \in E} a_{ij} \mathbf{v}_i^T \mathbf{v}_j : \|\mathbf{v}_1\|, \ldots, \|\mathbf{v}_n\| \leq 1 \right\}$$

(A small difference: for MAXCUT we had $\|\mathbf{v}_i\| = 1$, here we have ≤ 1; this will be technically more convenient.)

10.2.4 The Goemans–Williamson rounding by a random hyperplane will not work in general:

- The expected contribution of an edge $\{i, j\}$ after this rounding is within a constant factor of its contribution to $S_{\max}$. But now the edge contributions $a_{ij} x_i x_j$ can be both positive and negative, and so we cannot bound the approximation ratio.
- An analogy: If the price of every item on your shopping list changes by at most 5%, then the total also changes by no more than 5%; not a big deal. But if each of your monthly expenses changes by at most 5% and your monthly income also changes by no more than 5%, you may still have a serious problem with the monthly balance. (Also see under *economic crisis*.)

10.2.5 Indeed, MAXQP[G] is *harder* to approximate than MAXCUT:

- If $P \neq NP$, then no polynomial-time algorithm for $\text{MaxQP}[K_n]$ has a better approximation factor than $(\log n)^c$, for some small constant $c > 0$ (Arora et al. [ABH+05]).
- The integrality gap of the SDP relaxation of $\text{MaxQP}[K_n]$ is $\Omega(\log n)$ (Alon et al. [AMMN06], Khot and O'Donnell [KO09]).
- Assuming the Unique Games Conjecture (UGC), the problem $\text{MaxQP}[K_n]$ is hard to approximate with factor better than $\Omega(\log n)$ (Khot and O'Donnell [KO09]).

10.2.6 But there are impressive and beautiful *positive* results, which bound the integrality gap by a constant for a wide class of graphs, and give an efficient randomized rounding. We begin with a definition:

Let G be a (loopless) graph. The *Grothendieck constant K_G of G* is defined as

$$\sup \frac{S_{\max}}{\text{Opt}},$$

where Opt is the optimum value of $\text{MaxQP}[G]$, $S_{\max}$ is the optimum of the SDP relaxation, and the supremum is over all choices of the edge weights a_{ij} (not all zeros).

- So K_G is the largest integrality gap over all choices of edge weights.

10.2.7 It is not at all clear why K_G should be finite. We will cover the following strong upper bound.

Theorem (Alon et al. [AMMN06]). *For every graph G, we have*

$$K_G = O(\log \vartheta(\overline{G})),$$

where $\overline{G}$ is the complement of G and $\vartheta(.)$ is the Lovász theta function. Moreover, there is a randomized rounding algorithm which, for given G and weights a_{ij}, computes a solution of $\text{MaxQP}[G]$ with value at least $\Omega(S_{\max}/\log \vartheta(\overline{G}))$ in expected polynomial time.

10.2.8 Comments:

- We have $\vartheta(\overline{G}) \leq \chi(G)$, where $\chi(G)$ is the chromatic number (Theorem 3.7.2). Thus, if G is *bipartite* (as, e.g., in the CutNorm problem or in the Ising model for the cubic lattice), then there is a constant-factor approximation. Similarly if G has constant-bounded maximum degree, or it can be drawn on a fixed surface, or excludes a fixed minor...

- The fact that for the complete bipartite graph $K_{n,n}$ we have $K_{K_{n,n}} = O(1)$ is the *Grothendieck inequality* from the 1950s [Gro56]. This inequality inspired much of the work on the problem $\text{MaxQP}[G]$, including the theorem above.
- The classical *Grothendieck constant* is $\sup_n K_{K_{n,n}}$. The exact value is unknown; it is *in principle* computable up to any given error η (and only in principle so far; the running time is doubly exponential in η^{-1}); see Raghavendra and Steurer [RS09b].
- The UGC implies that $\text{MaxQP}[K_{n,n}]$ is hard to approximate within any constant smaller than the Grothendieck constant (see [RS09b]).
- **Open problem:** can we have $K_G << \log \vartheta(\overline{G})$? The known lower bound is only $\log \omega(G)$ (where $\omega(G)$ is the *clique number*).

10.3 The Rounding Algorithm

10.3.1 The rounding algorithm has several sources: Feige and Langberg [FL06], Charikar and Wirth [CW04], Nesterov [Nes98], Nemirovski, Roos and Terlaky [NRT99], Megretski [Meg01] ... Our presentation mostly follows K. Makarychev's thesis [Mak08].

10.3.2 The first idea of the rounding: Having computed vectors $\mathbf{v}_1, \ldots, \mathbf{v}_n$ attaining $S_{\max}$, we generate a random n-dimensional Gaussian $\boldsymbol{\gamma} \in \mathbb{R}^n$, and set $Z_i := \boldsymbol{\gamma}^T \mathbf{v}_i$.

10.3.3 We have

$$\mathbf{E}[Z_i Z_j] = \mathbf{v}_i^T \mathbf{v}_j.$$

Because:

- For *basis vectors*, $\boldsymbol{\gamma}^T \mathbf{e}_1, \ldots, \boldsymbol{\gamma}^T \mathbf{e}_n$ are independent $N(0,1)$ by definition, and so for $i \neq j$ we have $\mathbf{E}[(\boldsymbol{\gamma}^T \mathbf{e}_i)(\boldsymbol{\gamma}^T \mathbf{e}_j)] = 0$, while $\mathbf{E}[(\boldsymbol{\gamma}^T \mathbf{e}_i)^2]$ is the variance of an $N(0,1)$ random variable and thus equals 1. Hence $\mathbf{E}[(\boldsymbol{\gamma}^T \mathbf{e}_i)(\boldsymbol{\gamma}^T \mathbf{e}_j)] = \mathbf{e}_i^T \mathbf{e}_j$ for all i, j.
- For arbitrary $\mathbf{v}_i$ and $\mathbf{v}_j$, $\mathbf{E}[Z_i Z_j] = \mathbf{v}_i^T \mathbf{v}_j$ holds by bilinearity of the scalar product.

10.3.4 Therefore, $\mathbf{E}\left[\sum_{\{i,j\} \in E} a_{ij} Z_i Z_j\right] = S_{\max}$.

- We can thus "round" the vectors to numbers, but this is a big cheat, since with unbounded numbers we can make the quadratic form as large as we wish!

10.3.5 However, the Gaussians are mostly reasonably small. The actual algorithm: we scale the Z_i down by a suitable factor M, and then we truncate them to the interval $[-1, 1]$.

- Intuition: With quite high probability, we have $Z_i \in [-M, M]$ anyway, and then the rounding means just scaling the value of the solution down by M^2. So the rounding should produce a solution with value about $S_{\max}/M^2$.
- Of course, sometimes $Z_i \notin [-M, M]$, and this introduces additional error, which we have to deal with.

10.3.6 We define the "right" factor M, the smallest one for which the effect of the truncation is insignificant:

- Set

$$S_{\min} := \min\left\{ \sum_{\{i,j\}\in E} a_{ij}\mathbf{v}_i^T\mathbf{v}_j : \|\mathbf{v}_1\|,\ldots,\|\mathbf{v}_n\| \le 1 \right\}$$

(minimum of the semidefinite program for which $S_{\max}$ was the maximum; we have $S_{\min} < 0$, assuming that some $a_{ij} \ne 0$).
- Let $R := S_{\max} - S_{\min}$. So R is the *range* of the possible values of the semidefinite relaxation. (We note that $R \ge S_{\max}$.)
- Finally,

$$M := 3\sqrt{1 + \ln\frac{R}{S_{\max}}}$$

(here 3 is just a convenient constant).

10.3.7 The algorithm once again:

Randomized rounding for MAXQP[G]

1. Given $\mathbf{v}_1,\ldots,\mathbf{v}_n$ attaining $S_{\max}$, generate a random n-dimensional Gaussian $\boldsymbol{\gamma}$, and set $Z_i := \boldsymbol{\gamma}^T\mathbf{v}_i$, $i = 1, 2, \ldots, n$.
2. Compute R and M as above, and set

$$\tilde{Z}_i := \begin{cases} Z_i & \text{if } |Z_i| \le M \\ 0 & \text{otherwise.} \end{cases}$$

3. Return $x_i := \tilde{Z}_i/M$, $i = 1, 2, \ldots, n$.

Note that the a_{ij} enter the rounding only through M.

10.4 Estimating the Error

10.4.1 To estimate the expected value of the solution returned by the algorithm, we need to bound $\mathbf{E}\left[\sum_{\{i,j\}\in E} a_{ij}\tilde{Z}_i\tilde{Z}_j\right]$ from below.

10.4.2 Set $T_i := Z_i - \tilde{Z}_i$; then by linearity of expectation

$$\mathbf{E}\left[\sum_{\{i,j\}\in E} a_{ij}\tilde{Z}_i\tilde{Z}_j\right] = \sum_{\{i,j\}\in E} a_{ij}\Big(\mathbf{E}\left[Z_iZ_j\right] - \mathbf{E}\left[Z_iT_j\right] - \mathbf{E}\left[Z_jT_i\right] + \mathbf{E}\left[T_iT_j\right]\Big)$$

$$= S_{\max} - \mathbf{E}\left[\sum_{\{i,j\}\in E} a_{ij}(Z_iT_j + Z_jT_i)\right] + \mathbf{E}\left[\sum_{\{i,j\}\in E} a_{ij}T_iT_j\right].$$

10.4.3 Here is a general tool for dealing with the second and third terms.

Lemma. *Let* $X_1, X_2, \ldots, X_n$ *and* $Y_1, Y_2, \ldots, Y_n$ *be real random variables with* $\mathbf{E}\left[X_i^2\right] \leq A$ *and* $\mathbf{E}\left[Y_i^2\right] \leq B$ *for all* i *(no independence assumed). Then*

$$\mathbf{E}\left[\sum_{\{i,j\}\in E} a_{ij}(X_iY_j + X_jY_i)\right] \leq 2R\sqrt{AB},$$

with R *as in 10.3.6.*

10.4.4 First we show that if $\mathbf{E}\left[X_i^2\right] \leq 1$ for all i, then

$$S_{\min} \leq \mathbf{E}\left[\sum_{\{i,j\}\in E} a_{ij}X_iX_j\right] \leq S_{\max}.$$

This is where we connect random variables to vectors.

- Note that the right inequality holds with equality for $X_i := Z_i$, where $Z_i = \gamma^T\mathbf{v}_i$ as above. Similarly we can produce standard normal random variables giving equality on the left.
- The X_i all live on some probability space Ω. We consider the vector space of all real random variables X on Ω with $\mathbf{E}\left[X^2\right] < \infty$. In perhaps more familiar terms, this is the space $L^2(\Omega, \mu)$, where μ is the probability measure on Ω.
- A scalar product on this space is given by $\langle X, Y \rangle = \mathbf{E}\left[XY\right]$. It makes our space into a *Hilbert space* (a basic example of a Hilbert space; not hard to check the axioms).
- Any set of n vectors in a Hilbert space is isometric to a set of n vectors in $\mathbb{R}^n$. That is, in our case, there are vectors $\mathbf{v}_1, \ldots, \mathbf{v}_n \in \mathbb{R}^n$ with $\mathbf{v}_i^T\mathbf{v}_j = \mathbf{E}\left[X_iX_j\right]$, $i, j = 1, 2, \ldots, n$.

  ○ An alternative "direct" approach: One can check that the matrix $(\mathbf{E}\left[X_iX_j\right])_{i,j=1}^n$ is positive semidefinite, and construct the $\mathbf{v}_i$ by Cholesky factorization.

- Thus, $\mathbf{E}\left[\sum_{\{i,j\}\in E} a_{ij}X_iX_j\right] = \sum_{\{i,j\}\in E} a_{ij}\mathbf{v}_i^T\mathbf{v}_j \in [S_{\min}, S_{\max}]$.

10.4.5 Proof of Lemma 10.4.3:

- For the X_i and Y_i as in the lemma, we introduce the new variables

$$U_i := \frac{1}{2}\left(\frac{X_i}{\sqrt{A}} + \frac{Y_i}{\sqrt{B}}\right), \quad V_i := \frac{1}{2}\left(\frac{X_i}{\sqrt{A}} - \frac{Y_i}{\sqrt{B}}\right).$$

- Using $(x+y)^2 \le (x+y)^2 + (x-y)^2 = 2(x^2+y^2)$, we check that $\mathbf{E}\left[U_i^2\right] \le 1$, $\mathbf{E}\left[V_i^2\right] \le 1$. So we can apply the inequality in 10.4.4 to the U_i and V_i.
- Now

$$
\begin{aligned}
\mathbf{E}&\left[\sum_{\{i,j\}\in E} a_{ij}(X_iY_j + X_jY_i)\right] \\
&= 2\sqrt{AB}\left(\mathbf{E}\left[\sum_{\{i,j\}\in E} a_{ij}U_iU_j\right] - \mathbf{E}\left[\sum_{\{i,j\}\in E} a_{ij}V_iV_j\right]\right) \\
&\le 2\sqrt{AB}(S_{\max} - S_{\min}) = 2R\sqrt{AB}.
\end{aligned}
$$

The lemma is proved.　　□

10.4.6 In order to apply the lemma for estimating the error terms in the last line of 10.4.2, we need to bound $\mathbf{E}\left[Z_i^2\right]$ and $\mathbf{E}\left[T_i^2\right]$.

- Z_i is standard normal, so $\mathbf{E}\left[Z_i^2\right] = \mathrm{Var}\left[Z_i\right] = 1$, of course.
- For T_i, we have

$$A := \mathbf{E}\left[T_i^2\right] = \frac{2}{\sqrt{2\pi}}\int_M^\infty x^2 e^{-x^2/2}\,\mathrm{d}x.$$

- By the trick in 9.3.2, we can bound the integrand from above by $e^{-x^2/2}(x^2 + x^{-2})$, which can be integrated.
- Result:

$$\mathbf{E}\left[T_i^2\right] \le \sqrt{\frac{2}{\pi}}(M + \frac{1}{M})e^{-M^2/2} \le Me^{-M^2/2}$$

(since $M = 3\sqrt{1 + \ln(R/S_{\max})} \ge 3$, we have $M + \frac{1}{M} \le \frac{10}{9}M$).
- Estimating $M \le 3\sqrt{R/S_{\max}}$ (using $\ln x \le x - 1$), we get

$$A \le 3\sqrt{R/S_{\max}} \cdot e^{-9/2}\,(S_{\max}/R)^{9/2} < \frac{1}{10}\left(\frac{S_{\max}}{R}\right)^4 \le \frac{1}{10}\left(\frac{S_{\max}}{R}\right)^2.$$

10.4.7 Applying Lemma 10.4.3 to the second and third terms of the last line of 10.4.2 (once with $X_i = T_i$, $Y_i = Z_i$ and once with $X_i = -Y_i = T_i$) yields

$$
\begin{aligned}
\mathbf{E}\left[\sum_{\{i,j\}\in E} a_{ij}\tilde{Z}_i\tilde{Z}_j\right] &\ge S_{\max} - 2R\sqrt{A} - 2RA \\
&\ge S_{\max} - \frac{1}{5}S_{\max} - \frac{1}{5}S_{\max}\left(\frac{S_{\max}}{R}\right) \ge \frac{1}{2}S_{\max}.
\end{aligned}
$$

10.4.8 We can thus conclude:

> The expected value of the solution $x_1, \ldots, x_n$, i.e., of $\sum_{\{i,j\} \in E} a_{ij} x_i x_j$, is at least
> $$\frac{1}{2} \frac{S_{\max}}{M^2} \geq S_{\max} \cdot \Omega \left(\frac{1}{1 + \log \frac{R}{S_{\max}}} \right).$$

10.5 The Relation to $\vartheta(\overline{G})$

10.5.1 Here is the remaining part for the proof of Theorem 10.2.7.

Lemma. *For every graph G and for every choice of edge weights (not all zeros), we have*
$$\frac{R}{S_{\max}} \leq \vartheta(\overline{G}).$$

10.5.2 Proof:

- We go via a strict vector coloring.
- If $k := \vartheta(\overline{G})$, then there are unit vectors $\mathbf{u}_i$ with $\mathbf{u}_i^T \mathbf{u}_j = -\frac{1}{k-1}$ for every edge $\{i, j\}$ (this is a characterization of the Lovász theta function we had in Sect. 3.7).
- Let $\mathbf{v}_1, \ldots, \mathbf{v}_n$ be vectors attaining $S_{\min}$. Set $\mathbf{w}_i := \mathbf{u}_i \otimes \mathbf{v}_i$ (where $\otimes$ is the tensor product as in Sect. 3.4). Recalling the identity $(\mathbf{x} \otimes \mathbf{y})^T (\mathbf{x}' \otimes \mathbf{y}') = (\mathbf{x}^T \mathbf{x}')(\mathbf{y}^T \mathbf{y}')$, we see that $\|\mathbf{w}_i\| \leq 1$. Then

$$
\begin{aligned}
S_{\max} &\geq \sum_{\{i,j\} \in E} a_{ij} \mathbf{w}_i^T \mathbf{w}_j = \sum_{\{i,j\} \in E} a_{ij} (\mathbf{u}_i^T \mathbf{u}_j)(\mathbf{v}_i^T \mathbf{v}_j) \\
&= -\frac{1}{k-1} \sum_{\{i,j\} \in E} a_{ij} \mathbf{v}_i^T \mathbf{v}_j = \frac{-S_{\min}}{k-1}.
\end{aligned}
$$

Thus $\frac{R}{S_{\max}} = \frac{S_{\max} - S_{\min}}{S_{\max}} \leq k$, as the lemma claims. $\qquad\square$

10.5.3 Actually, for every G there are weights a_{ij} such that equality holds in the lemma (we will not prove this). So this is yet another characterization of $\vartheta(.)$.

Exercises

10.1 Consider the problem of correlation clustering as in 10.1.7. Prove that for every instance with n objects, there is a clustering with two clusters or with n clusters for which the correlation is at most three times the correlation for an optimal clustering.

10.2 Prove the two claims in 10.1.11 (those concerning $\|B\|_{\text{cut}}$).

10.3 (a) Let X be a positive semidefinite matrix, and let Y be the matrix with $y_{ij} = \arcsin x_{ij}$. Show that $Y - X \succeq 0$. (Hint: Consider the Taylor series of $\arcsin x - x$.)

(b) Let A be an $n \times n$ positive definite matrix, and consider the problem of maximizing $\mathbf{x}^T A \mathbf{x}$ over all $\mathbf{x} \in \{-1, 1\}^n$. We relax it to the vector program of maximizing $\sum_{i,j} a_{ij} \mathbf{v}_i^T \mathbf{v}_j$ subject to $\|\mathbf{v}_i\| = 1$ for all i (as in the Goemans–Williamson algorithm), we let $\mathbf{v}_1^*, \ldots, \mathbf{v}_n^*$ be an optimal solution of the latter, and let $\mathbf{y} \in \{-1, 1\}^n$ be obtained from the $\mathbf{v}_i^*$ by the random hyperplane rounding (again as in Goemans–Williamson). Prove, using (a), that the expected approximation ratio is at least $\frac{2}{\pi} \approx 0.636619\ldots$.

The result in (b) is due to Nesterov [Nes98]; most of the argument is also contained in Rietz [Rie74].

Chapter 11
Colorings with Low Discrepancy

11.1 Discrepancy of Set Systems

11.1.1 Let $V = \{1, 2, \ldots, n\}$ be a vertex set and let $\mathcal{F} = \{F_1, F_2, \ldots, F_m\}$ be a system of subsets of V. (We can also regard $(V, \mathcal{F})$ as a *hypergraph*.)

11.1.2 The basic problem in *combinatorial discrepancy theory* is to color each vertex $i \in V$ either red or blue, in such a way that each of the sets of $\mathcal{F}$ has roughly the same number of red points and blue points, as in the following schematic picture:

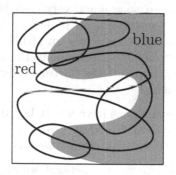

11.1.3 It is not always possible to achieve an exact splitting of each set. As an extreme example, if $\mathcal{F} := 2^V$ consists of all subsets of V, then there will always be a completely monochromatic set of size at least $\frac{n}{2}$.

11.1.4 The maximum deviation from an exact splitting, over all sets of $\mathcal{F}$, is the *discrepancy* of the set system $\mathcal{F}$. Formally, we let a *coloring* of $(V, \mathcal{F})$ be an arbitrary mapping $\chi: V \to \{-1, +1\}$.

B. Gärtner and J. Matoušek, *Approximation Algorithms and Semidefinite Programming*, DOI 10.1007/978-3-642-22015-9_11,

The *discrepancy* of $\mathcal{F}$ is

$$\mathrm{disc}(\mathcal{F}) := \min_{\chi} \mathrm{disc}(\mathcal{F}, \chi),$$

where the minimum is over all colorings $\chi\colon V \to \{-1, +1\}$, and

$$\mathrm{disc}(\mathcal{F}, \chi) := \max_{F \in \mathcal{F}} |\chi(F)|,$$

where we use the shorthand $\chi(F)$ for $\sum_{j \in F} \chi(j)$.

If $+1$'s are red and -1's are blue, then $\chi(F)$ is the number of red points in F minus the number of blue points in F, which we call the *imbalance* of F under χ.

11.1.5 Discrepancy has been investigated extensively, and there are many upper and lower bounds known for various set systems; see, e.g., [Mat10, Spe87, AS08] for introductions. There are also close connections to the classical subject of uniformly distributed point sets and sequences in various geometric domains, such as the unit square (see, e.g., [Mat10, ABC97]).

11.1.6 We will consider the algorithmic problem of computing a low-discrepancy coloring for a given set system $\mathcal{F}$.

- This question, in addition to its intrinsic interest, also presents a prototype question in the more general and very basic problem of *simultaneous rounding under linear constraints*. In such a problem, we have a vector $\mathbf{x} \in [-1, 1]^n$, which satisfies some system of linear constraints $A\mathbf{x} = \mathbf{b}$, and we would like to "round" each component to either $+1$ or -1 so that the resulting vector $\mathbf{z}$ almost satisfies the constraints, i.e., the vector $A\mathbf{z} - \mathbf{b}$ is small in a suitable sense.

  - In the case of discrepancy, we have $\mathbf{x} = \mathbf{0}$, $\mathbf{b} = \mathbf{0}$, and A is the incidence matrix of the set system $\mathcal{F}$ – that is, $a_{ij} = 1$ if $j \in F_i$, and $a_{ij} = 0$ otherwise. The rounding error is measured as $\|A\mathbf{x} - \mathbf{b}\|_\infty$.

- We cannot expect a satisfactory general solution of the simultaneous rounding problem, e.g., because of the hardness results for discrepancy mentioned below, but it is interesting to study special cases where a good rounding is possible.

11.1.7 Specifically, we will discuss a recent breakthrough by Bansal [Ban10], an algorithm for computing colorings with "reasonably low" discrepancy. It is based on semidefinite programming and it introduces a very interesting rounding strategy. The coloring is obtained using a random walk, driven by optimal solutions of suitable semidefinite programs – quite different from the "usual" hyperplane cuts.

11.1.8 First, $\mathrm{disc}(\mathcal{F})$ in itself is very hard to approximate. Charikar et al. [CNN11] proved that it is NP-hard to distinguish between set systems of discrepancy 0 and those with discrepancy of order $\sqrt{n}$.

- To appreciate this, one should know that $\mathrm{disc}(\mathcal{F}) = O(\sqrt{n \log(m/n)})$ for *every* system of $m \geq n$ sets on n points.
- It is easy to show that a *random* coloring has discrepancy $O(\sqrt{n \log m})$ with high probability, while the improvement of $\log m$ to $\log(m/n)$ is much harder; see [Mat10, Spe87, AS08].
- Thus, we have $\mathrm{disc}(\mathcal{F}) = O(\sqrt{n \log n})$ as long as m is bounded by a polynomial in n.

11.1.9 Discrepancy can be regarded as a measure of how complex a set system is. But it is not very well-behaved, since a system with almost the maximum possible discrepancy can be hidden in a system with zero discrepancy.

- Example: Take the complete set system $\mathcal{F} = 2^V$, make a disjoint copy V' of V, let F' be the clone of F in V'. The set system $(V \cup V', \{F \cup F' : F \in \mathcal{F}\})$ has discrepancy 0, yet we feel that it is as complex as $\mathcal{F}$.

11.1.10 A better behaved measure is the *hereditary discrepancy*, defined as

$$\mathrm{herdisc}(\mathcal{F}) := \max_{A \subseteq V} \mathrm{disc}(\mathcal{F}|_A).$$

Here $\mathcal{F}|_A$ denotes the *restriction* of the set system $\mathcal{F}$ to the ground set A, i.e., $\{F \cap A : F \in \mathcal{F}\}$.

- In other words, the enemy selects a subset $A \subseteq V$ and we must color its points red or blue so that each set in $\mathcal{F}$ is balanced; the uncolored points outside A do not count.
- Practically all known upper bounds on $\mathrm{disc}(\mathcal{F})$ also apply to $\mathrm{herdisc}(\mathcal{F})$.
- On the other hand, the question "Is $\mathrm{disc}(\mathcal{F}) \leq k$?" at least belongs to the class NP, while for "Is $\mathrm{herdisc}(\mathcal{F}) \leq k$?", membership in NP is open and maybe false.

11.1.11 Bansal's algorithm yields a coloring for a given $\mathcal{F}$ with discrepancy not much larger than $\mathrm{herdisc}(\mathcal{F})$:

Theorem (Bansal [Ban10]). *There is a randomized polynomial-time algorithm which, for an input set system $\mathcal{F}$ on n points, with m sets, and with hereditary discrepancy at most H, computes a coloring χ with* $\mathrm{disc}(\mathcal{F}, \chi) = O(H \log(mn))$.

- It may be tempting to conclude that the algorithm approximates the hereditary discrepancy with $O(\log(mn))$ factor, but this is not necessarily the case! Paradoxically, the algorithm may err on the good side; it may possibly compute a coloring with discrepancy much *smaller* than $\mathrm{herdisc}(\mathcal{F})$. (So we never learn that $\mathrm{herdisc}(\mathcal{F})$ is large.)

- Bansal's paper also has additional, technically subtler results, which save logarithmic factors in the discrepancy bound in certain special settings.

  o Our presentation of Bansal's algorithm below is somewhat simplified compared to his original formulation. However, for some of the additional results in his paper, our simplifications do not seem applicable.

11.1.12 Bansal's algorithm has converted several famous *existential* proofs in discrepancy theory into *constructive* ones.

- For example, Spencer [Spe85] proved in 1986 that every system of n sets on n points has discrepancy $O(\sqrt{n})$ (which is asymptotically tight in the worst case). The argument was existential. Bansal gave the first polynomial-time algorithm that finds a coloring with $O(\sqrt{n})$ discrepancy in this setting (this requires adding some twists to the algorithm presented here).
- As another example, the system of all *arithmetic progressions* on the ground set $\{1, 2, \ldots, n\}$ was known to have discrepancy of order $n^{1/4}$, but Bansal's algorithm is the first polynomial-time algorithm that can compute a coloring with discrepancy close to $n^{1/4}$ (with an extra logarithmic factor in this case).

11.2 Vector Discrepancy and Bansal's Random Walk Algorithm

11.2.1 First we set up a semidefinite relaxation of discrepancy. Instead of coloring by ± 1's, we color by unit vectors. We will thus talk about *vector discrepancy* vecdisc($\mathcal{F}$), which is the smallest $D \geq 0$ for which the following vector program is feasible:

$$
\left\| \sum_{j \in F_i} \mathbf{u}_j \right\|^2 \leq D^2, \quad i = 1, 2, \ldots, m,
$$
$$
\|\mathbf{u}_j\|^2 = 1, \quad j = 1, 2, \ldots, n.
$$

- This is indeed a relaxation of disc, so vecdisc($\mathcal{F}$) $\leq$ disc($\mathcal{F}$).
- We also introduce the *hereditary vector discrepancy* hervecdisc($\mathcal{F}$), as the maximum vector discrepancy of a restriction of $\mathcal{F}$ to a subset $A \subseteq V$.
- In the proof of Theorem 11.1.11, the algorithm will actually find a coloring with discrepancy at most $O(\text{hervecdisc}(\mathcal{F}) \log(mn))$.
- vecdisc($\mathcal{F}$) can be computed (up to a prescribed error) in polynomial time; for hervecdisc($\mathcal{F}$) we do not know.

11.2.2 In Bansal's algorithm we want to find a low-discrepancy coloring (by ± 1's). We will approach the desired coloring through a sequence of *semicolorings*, where a *semicoloring* is an arbitrary mapping $\xi \colon V \to [-1, 1]$.

- A semicoloring is like a coloring but by real numbers in $[-1, 1]$.
- The discrepancy $\text{disc}(\mathcal{F}, \xi)$ for a semicoloring is defined in the same way as for a coloring, as $\text{disc}(\mathcal{F}, \xi) = \max_{F \in \mathcal{F}} \left| \sum_{j \in F} \xi(j) \right|$.
- In the algorithm, we will represent a semicoloring by a point $\mathbf{x} \in [-1, 1]^n$.

11.2.3 Bansal's algorithm starts with the semicoloring $\mathbf{x}_0 := \mathbf{0}$, which has zero discrepancy but is rather useless as an "approximation" to a true coloring. Then it produces a sequence

$$\mathbf{x}_0, \mathbf{x}_1, \mathbf{x}_2, \ldots, \mathbf{x}_\ell \in [-1, 1]^n,$$

of semicolorings. Here ℓ, the length of the sequence, is a suitable parameter to be determined later.

- The algorithm is randomized.
- As we will see, with probability close to 1, the final semicoloring $\mathbf{x}_\ell$ is actually a coloring, i.e., all coordinates are ± 1's. This is the output of the algorithm.
- If $\mathbf{x}_\ell$ is not a coloring, we restart the algorithm from scratch.

11.2.4 The algorithm can be regarded as a random walk in the cube $[-1, 1]^n$. In the t-th step, $\mathbf{x}_t$ is obtained from $\mathbf{x}_{t-1}$ by a (small) random step, as follows:

- First we generate an increment $\boldsymbol{\Delta}_t \in \mathbb{R}^n$. It is random but chosen from a carefully crafted distribution; we will discuss this later.
- A "tentative value" of $\mathbf{x}_t$ is $\tilde{\mathbf{x}}_t := \mathbf{x}_{t-1} + \boldsymbol{\Delta}_t$. But we still need to truncate each coordinate to the interval $[-1, 1]$:

$$(\mathbf{x}_t)_j := \begin{cases} +1 & \text{if } (\tilde{\mathbf{x}}_t)_j \geq 1, \\ -1 & \text{if } (\tilde{\mathbf{x}}_t)_j \leq -1, \text{ and} \\ (\tilde{\mathbf{x}}_t)_j & \text{otherwise.} \end{cases}$$

- The increments $\boldsymbol{\Delta}_t$ are generated in such a way that once a coordinate of $\mathbf{x}_t$ reaches $+1$ or -1, it will never change. We can think of the faces of the cube as being "sticky"; as soon as the walk hits a face, it will stay in that face until the end.

  ○ More formally, we let $A_t := \{j \in V : (\mathbf{x}_{t-1})_j \neq \pm 1\}$ be the set of coordinates that are still *active* in the t-th step. We will make sure that $(\boldsymbol{\Delta}_t)_j = 0$ for all $j \notin A_t$.

- Figure 11.1 shows a schematic illustration of the random walk.

11.2.5 It remains to discuss how the increment $\boldsymbol{\Delta}_t$ is generated. The idea is that each (active) coordinate of $\boldsymbol{\Delta}_t$ is random, but the various coordinates are correlated so that the contribution of $\boldsymbol{\Delta}_t$ to the discrepancy is small.

- First, via semidefinite programming, we compute a coloring of the current active set A_t by unit vectors witnessing the vector discrepancy of the set system $\mathcal{F}|_{A_t}$.

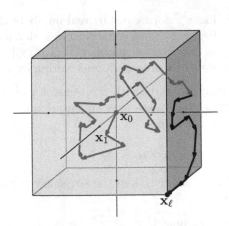

Fig. 11.1 Schematic illustration of the random walk

○ More explicitly, we compute unit vectors $\mathbf{u}_{t,j}$, $j \in A_t$, so that

$$\left\| \sum_{j \in F_i \cap A_t} \mathbf{u}_{t,j} \right\|^2 \le D^2$$

for all i, with $D \ge 0$ as small as possible.
○ For notational convenience, we also set $\mathbf{u}_{t,j} := \mathbf{0}$ for $j \notin A_t$.

• Next we generate a random vector $\boldsymbol{\gamma}_t$ from the n-dimensional standard normal (or Gaussian) distribution. That is, the coordinates of $\boldsymbol{\gamma}_t$ are independent $N(0, 1)$ random variables (also independent of $\boldsymbol{\gamma}_1, \ldots, \boldsymbol{\gamma}_{t-1}$).

○ Actually, the Gaussian distribution of the $\boldsymbol{\gamma}_t$ is not crucial for the algorithm. For example, independent random ± 1 vectors work as well, but the analysis becomes somewhat more complicated.

• Then we set
$$(\boldsymbol{\Delta}_t)_j := \sigma \boldsymbol{\gamma}_t^T \mathbf{u}_{t,j}, \quad j = 1, 2, \ldots, n.$$

Here σ is a sufficiently small parameter. As we will see in due time,

$$\sigma := \frac{1}{C_0 n \sqrt{\log n}},$$

with a sufficiently large constant C_0, will work (and a smaller σ, say n^{-2}, would work as well – only the running time would suffer).
• The length ℓ of the random walk should be set to $C_1 \sigma^{-2} \log n$, with another suitable constant C_1.

This concludes the description of Bansal's algorithm.

11.2.6 The idea why the algorithm works is outlined in the following two items (whose formulation is intentionally imprecise):

- First, the projection of the random walk to a given coordinate axis behaves like a one-dimensional random walk with increments having the normal distribution $N(0, \sigma^2)$. Such a walk will typically cross the boundary of the interval $[-1, 1]$ in about σ^{-2} steps, and after $\sigma^{-2} \log n$ steps it is even *very* likely to have crossed the boundary. But this means that $\mathbf{x}_\ell$ is typically a ± 1 vector.
- Second, the imbalance of a fixed set F_i starts out at 0 under the zero semicoloring $\mathbf{x}_0$, and in the t-th step it is changed by $\sum_{j \in F_i} \sigma \gamma_t^T \mathbf{u}_{t,j} = \sigma \gamma_t^T \mathbf{v}_{t,i}$, where $\mathbf{v}_{t,i} := \sum_{j \in F_i} \mathbf{u}_{t,j}$. But the $\mathbf{u}_{t,j}$ were selected with the goal of making all $\|\mathbf{v}_{t,i}\|$ small, and so the imbalance of each F_i grows only slowly during the algorithm.

For proving Theorem 11.1.11, we will establish formal counterparts of the two intuitive claims above, as follows.

11.2.7 Claim. *The algorithm produces a coloring with probability close to 1.*

We will do this in Sect. 11.3.

11.2.8 Claim. *With probability close to 1, the discrepancy of the resulting (semi)coloring is of order $O(H \log(mn))$, where H is the hereditary vector discrepancy of $\mathcal{F}$.*

We will undertake this in Sect. 11.4.

11.3 Coordinate Walks

11.3.1 Let $\mathbf{x}_0, \ldots, \mathbf{x}_\ell$ be the sequence generated by the algorithm, and let $j \in \{1, 2, \ldots, n\}$ be a fixed index. We call the sequence $(\mathbf{x}_0)_j, (\mathbf{x}_1)_j, \ldots, (\mathbf{x}_\ell)_j$ the j-th *coordinate random walk*.

11.3.2 We say that the j-th coordinate random walk *terminates* if $(\mathbf{x}_\ell)_j \in \{\pm 1\}$. To prove Claim 11.2.7, it suffices to show that, for every j, the probability that the j-th coordinate random walk does *not* terminate is at most n^{-2}. Then, with probability at least $1 - \frac{1}{n}$, all of the coordinate walks terminate (by the union bound).

- We note that this argument does not use any kind of independence among the coordinate walks. Necessarily so, since the whole point of the algorithm is that the coordinate walks are highly correlated!

11.3.3 To simplify notation, let us fix j and write $X_t := (\mathbf{x}_t)_j - (\mathbf{x}_{t-1})_j$, $t = 1, 2, \ldots, \ell$.

11.3.4 Let $t_0 \leq \ell$ be the last step of the j-th coordinate walk for which $(\mathbf{x}_{t_0})_j \in (-1, 1)$.

- By the rules of the algorithm, for $t \leq t_0$ we have $X_t = (\mathbf{\Delta}_t)_j = \sigma\gamma_t^T \mathbf{u}_{t,j}$ for some unit vector $\mathbf{u}_{t,j}$, where γ_t is n-dimensional Gaussian, independent of $\mathbf{u}_{t,j}$. Thus, as was mentioned in 9.3.3, X_t has the one-dimensional normal distribution $N(0, \sigma^2)$.
- We want to claim that $X_1, \ldots, X_{t_0}$ are independent random variables. But one has to be careful:

  - First, t_0 itself is not independent of $X_1, X_2, \ldots$, so even the formulation of such a claim may not be clear.
  - Moreover, the vector $\mathbf{u}_{t,j}$ depends on the previous history of the algorithm (more precisely, it depends on the set A_t, and through it on the earlier random choices made by the algorithm).

- Thus, we formulate our claim of independence in the following way. Let $Z_1', \ldots, Z_\ell'$ be a new sequence of independent random variables, each with the $N(0, \sigma^2)$ distribution, also independent of everything in the algorithm. We define another sequence $Z_1, Z_2, \ldots, Z_\ell$ of random variables by

$$Z_t := \begin{cases} X_t & \text{for } t \leq t_0 \\ Z_t' & \text{for } t > t_0. \end{cases}$$

We claim that $Z_1, \ldots, Z_\ell$ are independent.

  - Indeed, if we fix the values of $\gamma_1, \ldots, \gamma_{t-1}$ in the algorithm and also the values of the auxiliary variables $Z_1', \ldots, Z_{t-1}'$ arbitrarily, the values of $Z_1, \ldots, Z_{t-1}$ are determined uniquely, while Z_t has the $N(0, \sigma^2)$ distribution.
  - This easily implies the claimed independence of $Z_1, \ldots, Z_\ell$; the reader is invited to give a formal argument in Exercise 11.4.

11.3.5 According to the way the Z_t were defined, if the j-th coordinate walk does *not* terminate, then all the partial sums $Z_1 + Z_2 + \cdots + Z_t$ belong to $(-1, 1)$, $t = 1, 2, \ldots, \ell$. Thus we are left with the task of proving the following.

Lemma. *Let $Z_1, Z_2, \ldots, Z_\ell$ be independent random variables, each with the $N(0, \sigma^2)$ distribution. Then the probability that all of the partial sums $\sum_{i=1}^{t} Z_i$, $t = 1, 2, \ldots, \ell$, belong to the interval $(-1, 1)$ is at most $e^{-c_1 \lfloor \sigma^2 \ell \rfloor}$, for a suitable constant $c_1 > 0$.*

11.3.6 Here we can see the reason for choosing the walk length ℓ as we did, namely, $\ell := C_1 \sigma^{-2} \log n$. For this ℓ we get $e^{-c_1 \lfloor \sigma^2 \ell \rfloor} = e^{-c_1 \lfloor C_1 \log n \rfloor} \leq n^{-2}$ (for $C_1 := 3/c_1$, and n sufficiently large). So Claim 11.2.7 follows from the lemma.

11.3.7 It remains to prove the lemma. The asymptotic value of the probability in question is known quite precisely in the theory of random walks.

Here is a quick proof, which gives only a rough bound, but sufficient for our purposes.

- Let $k := \sigma^{-2}$ (assuming for convenience that this is an integer). Let us partition the sequence $Z_1, Z_2, \ldots$ into contiguous blocks of length k, and let S_j be the sum of the j-th block. Formally, $S_j := \sum_{i=(j-1)k+1}^{jk} Z_i$. The number of full blocks is $\lfloor \ell/k \rfloor$.
- **Fact:** if X, Y are independent $N(0, 1)$ random variables, and $a, b \in \mathbb{R}$, then $aX + bY \sim N(0, a^2 + b^2)$.

  - This is called the *2-stability* of the normal distribution.
  - **Sketch of a proof:** We may assume $a^2 + b^2 = 1$ (re-scaling). The vector (X, Y) has the 2-dimensional standard normal distribution, rotationally symmetric. Thus, its scalar product with an arbitrary unit vector has the 1-dimensional $N(0, 1)$ distribution. It remains to observe that $aX + bY$ is the scalar product of (X, Y) with (a, b).

- So each S_j has the standard normal distribution $N(0, k\sigma^2) = N(0, 1)$. Thus, $\text{Prob}[|S_j| \geq 2] \geq c_0$ for a suitable positive c_0 (by looking at a table of the normal distribution we can find that $c_0 \approx 0.0455$).
- If $\sum_{i=1}^{t} Z_i \in (-1, 1)$ for all $t = 1, 2, \ldots, \ell$, then necessarily $|S_j| < 2$ for all j. The S_j are independent, and thus the probability of the latter is at most $(1 - c_0)^{\lfloor \ell/k \rfloor} = e^{-c_1 \lfloor \sigma^2 \ell \rfloor}$. The lemma is proved, and so is Claim 11.2.7.

11.4 Set Walks

11.4.1 It remains to prove Claim 11.2.8; concretely, we will prove

$$\text{Prob}[\text{disc}(\mathcal{F}, \mathbf{x}_\ell) > D_{\max}] \leq \frac{1}{n},$$

where $D_{\max} = O(H \log(mn))$ is the desired bound on the discrepancy.

11.4.2 Let us fix a set $F_i \in \mathcal{F}$, and let $D_i := \sum_{j \in F_i} (\mathbf{x}_\ell)_j$ be its imbalance in the final (semi)coloring $\mathbf{x}_\ell$. We will prove that $\text{Prob}[|D_i| > D_{\max}] \leq \frac{1}{mn}$ for every i, and Claim 11.2.8 will follow by the union bound.

11.4.3 We recall how the j-th coordinate of the current semicoloring $\mathbf{x}_t$ develops as t goes from 0 to ℓ.

- It starts with $(\mathbf{x}_0)_j = 0$, then it changes by the random increments $(\mathbf{\Delta}_t)_j$, $t = 1, 2, \ldots, t_0$, then at some step $t_0 + 1$ it gets truncated to $+1$ or -1, and then it stays fixed until the end (it may also happen, with some small probability, that $t_0 = \ell$ and no truncation occurs).
- Since $(\mathbf{\Delta}_t)_j = 0$ for $t > t_0 + 1$, we can write

$$(\mathbf{x}_\ell)_j = \sum_{t=1}^{\ell}(\mathbf{\Delta}_t)_j + T_j,$$

where T_j is a "truncation effect," reflecting the fact that $(\mathbf{x}_{t_0+1})_j$ equals ± 1 and not $(\mathbf{x}_{t_0} + \mathbf{\Delta}_{t_0+1})_j$.

- We have $|T_j| \le |(\mathbf{\Delta}_{t_0+1})_j|$, and as we know, $(\mathbf{\Delta}_{t_0+1})_j \sim N(0, \sigma^2)$.
- Here is where our choice $\sigma := 1/(C_0 n\sqrt{\log n})$ comes from: we will now show that for σ this small, all truncation effects are negligible with probability close to 1. Quantitatively, we claim that, for each j,

$$\mathrm{Prob}\left[|T_j| > \tfrac{1}{n}\right] \le \tfrac{1}{n^3}$$

(both $\frac{1}{n}$ and $\frac{1}{n^3}$ are chosen somewhat arbitrarily here; we could as well take $\frac{1}{n^{10}}$).

- **Proof:**

  o We just employ a tail bound for the standard normal distribution, e.g., Lemma 9.3.2. For a standard normal random variable Z, that formula gives $\mathrm{Prob}\left[|Z| \ge \lambda\right] \le e^{-\lambda^2/2}$ for all $\lambda \ge 1$.
  o In our situation,

$$\mathrm{Prob}\left[|T_j| > \frac{1}{n}\right] \le \mathrm{Prob}\left[|\sigma Z| \ge \frac{1}{n}\right] = \mathrm{Prob}\left[|Z| \ge \frac{1}{\sigma n}\right]$$

$$\le e^{-\sigma^{-2}n^{-2}/2} = e^{-(C_0^2 \log n)/2} \le \frac{1}{n^3}$$

for a suitable C_0.

11.4.4 Thus, with probability at least $1 - \frac{1}{n^2}$, the total contribution of the truncation effects T_j to the discrepancy of each set F_i is at most 1. So instead of D_i, it suffices to bound the "pure random walk" quantity

$$\tilde{D}_i := \sum_{j \in F_i}\sum_{t=1}^{\ell}(\mathbf{\Delta}_t)_j = \sum_{t=1}^{\ell}\sum_{j \in F_i}(\mathbf{\Delta}_t)_j$$

$$= \sum_{t=1}^{\ell}\sum_{j \in F_i}\sigma\gamma_t^T\mathbf{u}_{t,j} = \sum_{t=1}^{\ell}\sigma\gamma_t^T\mathbf{v}_{t,i},$$

where $\mathbf{v}_{t,i} = \sum_{j \in F_i}\mathbf{u}_{t,j}$.

11.4.5 Now, finally, the careful choice of the $\mathbf{u}_{t,j}$ (see 11.2.5) comes into play: we know that $\|\mathbf{v}_{t,i}\| \le H$ for all t and i. Writing $Y_t := \sigma\gamma_t^T\mathbf{v}_{t,i}$ (we recall that i is considered fixed), we get that $Y_t \sim N(0, \beta_t^2)$, where $0 \le \beta_t \le \sigma H$.

- Intuitively, things should be simple here: the sum $\tilde{D}_i = Y_1 + \cdots + Y_\ell$ should have a distribution like $N(0, \ell\sigma^2 H^2)$, and hence we should get an exponential tail bound; the probability of $|\tilde{D}_i|$ exceeding $\sigma H \sqrt{\ell}$, the standard deviation, λ times, should behave like $e^{-\lambda^2/2}$.

- However, unlike in the case of the coordinate walks, we cannot claim that the Y_t are independent, because the variance β_t^2 of Y_t does depend on the random choices made by the algorithm in step 1 through $t-1$.
- So we need a more sophisticated and technical tool, such as the following lemma.

11.4.6 Lemma. *Let $W_1, \ldots, W_\ell$ be independent random variables on some probability space, and let Y_t be a function of $W_1, \ldots, W_t$, $t = 1, 2, \ldots, \ell$. Suppose that, conditioned on $W_1, \ldots, W_{t-1}$ attaining some arbitrary values $w_1, \ldots, w_{t-1}$, the distribution of Y_t is $N(0, \beta_t^2)$, where β_t may depend on $w_1, \ldots, w_{t-1}$, but we always have $\beta_t \leq \beta$. Then $Y := Y_1 + \cdots + Y_\ell$ satisfies the tail bound*

$$\mathrm{Prob}\left[|Y| > \lambda \beta \sqrt{\ell}\right] \leq 2e^{-\lambda^2/2}, \quad \text{for all } \lambda \geq 0.$$

- We will use the lemma with $W_t := \gamma_t$. Our λ is $\frac{D_{\max}}{\sigma H \sqrt{\ell}}$, where $D_{\max} = C_2 H \log(mn)$, $\sigma = 1/(C_0 n \sqrt{\log n},)$, and $\ell = C_1 \sigma^{-2} \log n$. We thus calculate that $\lambda \geq C_3 \sqrt{\log(mn)}$, with a constant C_3 that can be made as large as needed by adjusting the constant C_2 from $D_{\max}$. Then

$$\mathrm{Prob}\left[|\tilde{D}_i| > D_{\max}\right] \leq 2e^{-\lambda^2/2} \leq \frac{1}{n^2 m^2},$$

and, as announced, Claim 11.2.8 follows by the union bound.

- Some readers may have recognized that we are really talking about a *martingale* in the lemma (while those not familiar with martingales may ignore this remark). They may also know *Azuma's inequality*, which gives a tail bound for martingales. That inequality assumes pointwise bounded martingale differences, and thus is not directly applicable in our setting. However, its *proof* is applicable with only a minor adjustment, and this is how we prove the lemma.

11.4.7 Proof of the lemma:

- We will prove the upper tail bound, $\mathrm{Prob}\left[Y > \lambda \beta \sqrt{\ell}\right] \leq e^{-\lambda^2/2}$; the lower one follows by applying the upper tail to $-Y$.
- As in the usual proof of the Chernoff inequality, the main trick is to bound the *moment generating function* $G(u) := \mathbf{E}\left[e^{uY}\right]$, where u is a real parameter.
- By induction on t, we will show that $\mathbf{E}\left[e^{u(Y_1+\cdots+Y_t)}\right] \leq e^{u^2 t \beta^2/2}$ (the basis is for $t = 0$).
- The expectation is over a random choice of $W_1, \ldots, W_t$. We can do the expectation over W_t first, regarding $W_1, \ldots, W_{t-1}$ fixed, and then the expectation of the result over $W_1, \ldots, W_{t-1}$ (this is Fubini's theorem).

- For $W_1, \ldots, W_{t-1}$ fixed, $Y_1, \ldots, Y_{t-1}$ are fixed, and so

$$
\begin{aligned}
\mathbf{E}_{W_t}\left[e^{u(Y_1 + \cdots + Y_t)}\right] &= e^{u(Y_1 + \cdots + Y_{t-1})}\mathbf{E}_{W_t}\left[e^{uY_t}\right] \\
&= e^{u(Y_1 + \cdots + Y_{t-1})}e^{u^2\beta_t^2/2} \le e^{u^2\beta^2/2}e^{u(Y_1 + \cdots + Y_{t-1})}.
\end{aligned}
$$

  ○ Here we have used that, with $W_1, \ldots, W_{t-1}$ fixed, we have $Y_t \sim N(0, \beta_t^2)$. Then $\mathbf{E}\left[e^{uY_t}\right] = e^{u^2\beta_t^2/2}$ is a standard fact about the normal distribution, which is also easy to check (a substitution converts the required integral to $\int_{-\infty}^{\infty} e^{-x^2/2}\mathrm{d}x$).

- Now we take expectation over $W_1, \ldots, W_{t-1}$ and use the inductive hypothesis, and thus finish the inductive step. We have shown that $\mathbf{E}\left[e^{uY}\right] \le e^{u^2\ell\beta^2/2}$.

- Next we use the Markov inequality for the random variable e^{uY}:

$$
\begin{aligned}
\mathrm{Prob}\left[Y > \lambda\beta\sqrt{\ell}\right] &= \mathrm{Prob}\left[e^{uY} > e^{u\lambda\beta\sqrt{\ell}}\right] \\
&\le \mathbf{E}\left[e^{uY}\right]/e^{u\lambda\beta\sqrt{\ell}} \le e^{u^2\ell\beta^2/2 - u\lambda\beta\sqrt{\ell}}.
\end{aligned}
$$

Substituting $u := \lambda/(\beta\sqrt{\ell})$ gives the required tail bound, and the lemma is proved. So are Claim 11.2.8 and Theorem 11.1.11.

Exercises

11.1 (a) Show that every set system on n points has vector discrepancy at most $\sqrt{n}$.
(b) Show that this bound is tight, possibly up to a multiplicative constant independent of n.
(c) Let $\mathcal{F}$ be a system of n sets on a set V of points, such that every point is contained in exactly r sets of $\mathcal{F}$, and for every two distinct points $i, j \in V$, there are exactly t sets $F \in \mathcal{F}$ with $\{i, j\} \subseteq F$. Prove that $\mathrm{vecdisc}(\mathcal{F}) \ge \sqrt{r - t}$.

   Hint: Instead of $\max_{F \in \mathcal{F}}\left\|\sum_{j \in F} \mathbf{u}_j\right\|^2$, estimate $\sum_{F \in \mathcal{F}}\left\|\sum_{j \in F} \mathbf{u}_j\right\|^2$.

Remark: The so-called *Hadamard designs* provide set systems as above with $n = 4m + 3$, $r = 2m + 1$, and $t = m$ for infinitely many values of m, and thus they show that a system of n sets on n points can have vector discrepancy $\Omega(\sqrt{n})$.

11.2 Use the probabilistic method (and, in particular, the Chernoff bound for the sum of independent ± 1 random variables) to show that $\mathrm{disc}(\mathcal{F}) = O(\sqrt{n \log m})$ for every system of m sets on n points.

11.3 Use the probabilistic method to show the existence of set systems with n^2 sets on n points and with discrepancy $\Omega(\sqrt{n \log n})$. (Together with Exercise 11.1(a) this shows that the gap between vecdisc and disc can be at least of order $\sqrt{\log m}$, for $m = n^2$. The complete set system 2^V exhibits a similar gap for $m = 2^n$.)

Hint: Fix an arbitrary coloring χ and show that the discrepancy of χ for a system of n^2 independent random sets is below $c\sqrt{n \log n}$ with probability smaller than 2^{-n}. A random set is obtained by including each point independently with probability $\frac{1}{2}$.

11.4 Let $Z_1, Z_2, \ldots, Z_n$ be real random variables, and suppose that there are distributions $\mathcal{D}_1, \ldots, \mathcal{D}_n$ such that for each $t = 1, 2, \ldots, n-1$ and for every $z_1, \ldots, z_{t-1} \in \mathbb{R}$, the conditional distribution of Z_t given that $Z_1 = z_1, \ldots, Z_{t-1} = z_{t-1}$ is $\mathcal{D}_t$. Prove rigorously that $Z_1, \ldots, Z_t$ are independent. (Recall that real random variables $X_1, \ldots, X_n$ are independent if for every index set $I \subseteq \{1, 2, \ldots, n\}$ and for every $a_1, \ldots, a_n \in \mathbb{R}$, we have $\mathrm{Prob}[X_i \leq a_i \text{ for all } i \in I] = \prod_{i \in I} \mathrm{Prob}[X_i \leq a_i]$.)

Chapter 12
Constraint Satisfaction Problems, and Relaxing Them Semidefinitely

12.1 Introduction

12.1.1 Constraint satisfaction problems constitute an important and much studied class of computational problems.

We will discuss them from the point of view of SDP-based approximation algorithms. They are also investigated from other angles, each of them giving rise to a major field of research. We will mention some of these other aspects only very briefly and omit others completely.

12.1.2 Before defining a constraint satisfaction problem in general, we recall a few things about MAX-3-SAT and similar computational problems.

The reader is probably familiar with 3-SAT, a prototype NP-complete problem.

- The input to 3-SAT is a Boolean formula in conjunctive normal form with three literals per clause, such as

$$(x_1 \vee x_2 \vee \overline{x}_3) \wedge (x_1 \vee x_3 \vee x_5) \wedge (\overline{x}_2 \vee \overline{x}_4 \vee x_5) \wedge (x_2 \vee \overline{x}_3 \vee x_5).$$

- There are n Boolean variables $x_1, \ldots, x_n$, and m clauses, where each clause is the disjunction of three or fewer literals (or we can insist on exactly three literals, since the same literal can be repeated twice or three times in a clause).
- A literal can be x_i or $\overline{x}_i$, where $\overline{x}_i$ is the negation of x_i.
- $\wedge$ means "and" (conjunction), and $\vee$ means "or" (disjunction).
- The goal in 3-SAT is to find an assignment of truth values (True or False) to the x_i such that all of the clauses are satisfied.

12.1.3 MAX-3-SAT is an optimization version of 3-SAT; the goal is to find an assignment satisfying as many clauses of a given formula as possible.

- We recall that an α-approximation algorithm for MAX-3-SAT is an algorithm which, for every input formula φ, computes an assignment satisfying

B. Gärtner and J. Matoušek, *Approximation Algorithms and Semidefinite Programming*, DOI 10.1007/978-3-642-22015-9_12,
© Springer-Verlag Berlin Heidelberg 2012

at least $\alpha \cdot$ Opt of the clauses, where Opt is the maximum number of satisfiable clauses.

- One of the most spectacular results in the theory of approximation is: The best approximation factor for MAX-3-SAT achievable by any polynomial-time algorithm is $\frac{7}{8}$, assuming $P \neq NP$.
- The *upper bound*, inapproximability above $\frac{7}{8}$ [Hås01], is a great achievement of the theory related to the PCP theorem, beyond our scope.
- The *lower bound*, i.e., a $\frac{7}{8}$-approximation algorithm, is easy for formulas where every clause involves literals with *three distinct variables*.

  ○ Indeed, let us consider, e.g., the clause $x_1 \vee \overline{x}_2 \vee x_3$. There are 8 possible assignments to x_1, x_2, x_3, and 7 of them satisfy the clause – the only exception is (False, True, False). Similarly, any other clause involving 3 distinct variables is satisfied by $\frac{7}{8}$ of the possible assignments.

  ○ Hence, if we select a random assignment, where each x_i attains the values True and False with probability $\frac{1}{2}$, independent of all the other variables, then each of the clauses is satisfied with probability $\frac{7}{8}$. So the expected number of clauses satisfied by a random assignment is $\frac{7}{8}m$.

  ○ This can be turned into a randomized algorithm that finds, for every input formula with the above restriction, an assignment satisfying at least $\lfloor \frac{7}{8}m \rfloor$ clauses (Exercise 12.1).

- Now the clauses with only 1 or 2 literals (or with a multiple occurrence of the same variable) may look like only a minor nuisance. But a $\frac{7}{8}$-approximation algorithm for MAX-3-SAT with such clauses allowed is *much* harder [KZ97] (actually, the approximation ratio has not even been proved rigorously – there is just a strong numerical evidence). It is based on a sophisticated semidefinite relaxation, a clever rounding method, and a very complicated computer-assisted analysis.

12.1.4 One can also consider **MAX-2-SAT** (two literals per clause), and in general MAX-k-SAT. An interesting feature of the $k = 2$ case is that the satisfiability version (2-SAT) is known to be decidable in polynomial time, yet MAX-2-SAT is NP-hard (and hard to approximate with factor better than roughly 0.954 [Hås01]).

12.2 Constraint Satisfaction Problems

12.2.1 A clause in a 3-SAT formula, say $x_1 \vee x_2 \vee x_3$, puts a constraint on the possible values of x_1, x_2, x_3: This particular clause requires that at least one of them should be true. In various real-life or other problems, we often want to express different kinds of constraints, e.g., that exactly one of x_1, x_2, x_3 should be true, or all of them should be true, etc.

12.2.2 The 3-SAT formulas are universal, i.e., every Boolean function can be expressed by such a formula, so in principle, we can translate any kind of constraints into a 3-SAT formula.

But such translation may be nonintuitive and produce large formulas. More seriously, since 3-SAT formulas are so powerful, they are also hard to deal with, while for formulas built from other, more special classes of constraints we can sometimes get better approximability or even polynomial-time algorithms.

12.2.3 A *constraint satisfaction problem* can be regarded as a generalization of k-SAT where clauses are replaced by arbitrary predicates in k variables, and the variables may generally attain more than two values.

- The variables x_i attain values in a finite domain D. We will mostly stick to the Boolean setting with $D = \{\text{False}, \text{True}\}$. (More generally, each x_i might have its own domain D_i.)
- A k-*ary predicate* over D is a Boolean function $P\colon D^k \to \{\text{False}, \text{True}\}$. It evaluates to True on the k-tuples for which the predicate is satisfied.

  ○ For example, the clause $x_1 \vee \overline{x}_2 \vee x_3$ can be regarded as the ternary predicate giving False on the triple $(\text{False}, \text{True}, \text{False})$ and True on the 7 remaining triples in $\{\text{False}, \text{True}\}^3$.

  ○ We will often write predicates less formally; e.g., $x \neq y$ means the binary predicate $P(x, y)$ with $P(a, b) = \text{True}$ for all $a, b \in D$ with $a \neq b$ and $P(a, a) = \text{False}$ for all $a \in D$.

- A class of constraint satisfaction problems is specified by k, D, and a list $\mathcal{P}$ of k-ary predicates that can appear in the instances. We write k-CSP$[\mathcal{P}]$ for such a class.

  ○ For example, for 3-SAT we have $k = 3$, $D = \{\text{False}, \text{True}\}$, and $\mathcal{P}$ consists of all ternary predicates that give False for exactly one triple of values of the variables.

- An instance of a 3-CSP$[\mathcal{P}]$ may look like this, for example:

$$P_1(x_1, x_2, x_3) \wedge P_2(x_2, x_1, x_4) \wedge \cdots \wedge P_m(x_3, x_5, x_1),$$

$P_1, P_2, \ldots, P_m \in \mathcal{P}$.
- The terms $P_1(x_1, x_2, x_3)$, $P_2(x_2, x_1, x_4)$, etc., are usually called the *constraints* of the instance. But here we will avoid that word, because we already have constraints in semidefinite programs. If necessary, we will call them *generalized clauses* or simply *clauses*.

12.2.4 Here is a summary of the general definition.

A *class of constraint satisfaction problems* k-CSP$[\mathcal{P}]$ is specified by

- A finite domain D
- A natural number k (the arity)
- A set $\mathcal{P}$ of k-ary predicates over D

k and $|D|$ are usually treated as *constants*. An *instance* of k-CSP$[\mathcal{P}]$ is

$$P_1(x_{i_{11}}, x_{i_{12}}, \ldots, x_{i_{1k}}) \wedge P_2(x_{i_{21}}, x_{i_{22}}, \ldots, x_{i_{2k}}) \wedge \cdots$$

$$\wedge P_m(x_{i_{m1}}, x_{i_{m2}}, \ldots, x_{i_{mk}}),$$

where $P_1, \ldots, P_m \in \mathcal{P}$ and $i_{11}, i_{12}, \ldots, i_{mk} \in \{1, 2, \ldots, n\}$. An *assignment* for this instance is an n-tuple $(a_1, \ldots, a_n) \in D^n$, and the generalized clause $P_\ell(x_{i_{\ell 1}}, \ldots, x_{i_{\ell k}})$ is *satisfied* by that assignment if $P_\ell(a_{i_{\ell 1}}, \ldots, a_{i_{\ell k}}) = \mathsf{True}$.

We also define the optimization version **Max-k-CSP$[\mathcal{P}]$**, where the goal is to satisfy as many generalized clauses as possible.

12.2.5 Here is a sample of binary constraint satisfaction problems.

- For $D = \{\mathsf{False}, \mathsf{True}\}$ and $\mathcal{P}$ consisting of the single predicate "$x \neq y$," the problem Max-2-CSP$[\mathcal{P}]$ encodes MaxCut. The variables $x_1, \ldots, x_n$ correspond to vertices of the given graph G, and for every edge $\{i, j\} \in E(G)$, we introduce the clause $x_i \neq x_j$. Then the number of clauses satisfied by an assignment is the number of edges in the cut defined by the set $\{i : x_i = \mathsf{True}\}$.
- With the predicate "$x \wedge \overline{y}$," we can encode MaxDiCut, the problem of finding a maximum directed cut in a directed graph (i.e., we look for a subset $A \subseteq V(G)$ such that the maximum possible number of directed edges go from A to $V(G) \setminus A$).
- We already know Max-2-Sat. Its $\mathcal{P}$ consists of four binary predicates.
- Max-2-And, with the predicates "$x \wedge y$", "$\overline{x} \wedge y$", "$x \wedge \overline{y}$", and "$\overline{x} \wedge \overline{y}$", is conjectured to have the worst approximation ratio among all Boolean 2-CSP; see Austrin [Aus07]. Curiously, the corresponding satisfiability problem is trivial.
- A very trivial example is obtained for the predicate "$x = y$."
- As a non-Boolean example, we have graph 3-coloring, where we set $D := \{1, 2, 3\}$ and the predicate is "$x \neq y$." Clauses correspond to edges.

12.2.6 Comments:

- Constraint satisfaction problems can be further generalized. For example, each clause may have a nonnegative real weight, and we want to maximize the total weight of the satisfied clauses. Or, instead of predicates

$P: D^k \to \{\text{False}, \text{True}\}$, we can consider nonnegative real *payoff functions* $P: D^k \to [0, \infty)$, which assign a payoff to each possible setting of the k variables, and we want to maximize the total payoff. The methods discussed below can easily be generalized to this payoff setting.

- Another useful view of CSP's is through *homomorphisms*. Every CSP can be reformulated as follows: Given an input object, decide whether it admits a homomorphism into a fixed target object. Here the objects can be, e.g., directed graphs, or more generally, colored hypergraphs. Here we will not pursue this direction.

- Various heuristic approaches to solving CSP's in practice constitute a major research field.

- A great effort has been put into classifying the CSP's according to their computational complexity, using tools from logic, model theory, and algebra. A fundamental open problem in the field is the *dichotomy conjecture* (Feder and Vardi [FV98]), asserting that every class k-CSP[$\mathcal{P}$], considered as a decision problem in the obvious way, is either in P (polynomial-time decidable), or NP-complete. The conjecture has been verified for domain size at most 3 (Bulatov [Bul02]), and many other partial results are known.

- Another ongoing project is to classify the problems MAX-k-CSP[$\mathcal{P}$] according to the best achievable approximation factors. Here semidefinite programming is an indispensable tool for lower bounds (approximation algorithms), while inapproximability results rely on PCP techniques and the Unique Games Conjecture. Surprisingly, the inapproximability proofs under the UGC are closely related to the geometry of certain semidefinite relaxations of the considered CSP's, and to their integrality gaps (see, e.g., [KKMO07, Aus07, Rag08, Rag09]).

The Unique Games Conjecture for Us Laymen, Part II

12.2.7 Now we can put the Unique Games Conjecture in a wider context.

- For constraint satisfaction problem MAX-k-CSP[$\mathcal{P}$] we can consider a $(1 - \delta, \varepsilon)$-gap version (distinguish the instances where at least a $(1 - \delta)$-fraction of the clauses can be satisfied from those where no more than an ε-fraction are satisfiable).

- Let $\mathcal{P}_{\text{perm}}$ denote the set of all binary predicates over some given domain D that correspond to *permutations*. That is, for every $a \in D$ there is exactly one $b \in D$ with $P(a, b) = \text{True}$, and also exactly one $b \in D$ with $P(b, a) = \text{True}$.

- Here is a different way of stating the UGC (the formulation introduced by Khot [Kho02]): Assuming $P \neq NP$, for every $\varepsilon > 0$ there exists an integer q such that the $(1 - \varepsilon, \varepsilon)$-gap version of MAX-2-CSP[$\mathcal{P}_{\text{perm}}$] over a q-element domain D cannot be solved in polynomial time.

12.2.8 Comments:

- The problem MAX-2-LIN$(\bmod q)$ introduced in 8.5.3 is a special case of MAX-2-CSP$[\mathcal{P}_{\text{perm}}]$ (with domain size q). Indeed, if we have a linear equation $x-y = c \,(\bmod q)$, then for every possible value of x there exists exactly one value of y satisfying the equation, and thus such an equation corresponds to a predicate in $\mathcal{P}_{\text{perm}}$.
- However, only a small subset of all possible predicates in $\mathcal{P}_{\text{perm}}$ can be obtained in this way. So MAX-2-LIN$(\bmod q)$ might potentially be easier than MAX-2-CSP$[\mathcal{P}_{\text{perm}}]$. But it is not: As was proved by Khot et al. [KKMO07], the UGC as stated above, with MAX-2-CSP$[\mathcal{P}_{\text{perm}}]$, implies the version with MAX-2-LIN$(\bmod q)$ as in Sect. 8.5.
- It is easy to see that the basic satisfiability version 2-CSP$[\mathcal{P}_{\text{perm}}]$ is polynomially solvable.
- The example of MAXCUT shows that MAX-2-CSP$[\mathcal{P}_{\text{perm}}]$ is NP-hard.
- The subexponential algorithm of Arora et al. [ABS10] mentioned in 8.5.5 applies to every MAX-2-CSP$[\mathcal{P}_{\text{perm}}]$.

12.3 Semidefinite Relaxations of 2-CSP's

12.3.1 First we will set up a semidefinite relaxation of MAX-2-SAT. We replace each Boolean variable x_i by a vector variable.

- There are two main ways of "encoding" the truth values by a vector in the literature. They are equivalent but they look rather different. We will mainly work with the one which is more suitable for generalizations to non-Boolean domains, and perhaps more intuitive. However, the other one seems prevalent in the literature, and we will mention it as well.

12.3.2 We begin with formulating MAX-2-SAT as a quadratic program.

- We represent each x_i by a real variable y_i, where $y_i = 1$ corresponds to $x_i = \textsf{True}$ and $y_i = 0$ to $x_i = \textsf{False}$.
- Instead of requiring $y_i \in \{0,1\}$ explicitly, we express it by the quadratic constraint $y_i(1 - y_i) = 0$.
- The clauses are reflected in the objective function. We want to maximize $\sum_{\ell=1}^{m} f_\ell(y_1, \ldots, y_n)$. If, for example, the first clause is $x_1 \vee x_2$, then

$$f_1(y_1, \ldots, y_n) := y_1 y_2 + y_1(1 - y_2) + (1 - y_1)y_2.$$

(This could be simplified to $y_1 + y_2 - y_1 y_2$ but for now we prefer to leave it as is.) The three terms on the r.h.s. correspond to the three possible assignments to x_1 and x_2 that satisfy the clause: $y_1 y_2$ gives 1 for $y_1 = y_2 = 1$ and 0 otherwise, $y_1(1 - y_2)$ gives 1 for $y_1 = 1$, $y_2 = 0$ and 0 otherwise, etc. Thus, the objective function expresses the number of satisfied clauses.

12.3.3 Having set up the quadratic program, we want to relax it by replacing each y_i by a vector $\mathbf{t}_i$. (We use the letter $\mathbf{t}$ instead of the usual $\mathbf{v}$ – this should remind us that $\mathbf{t}_i$ expresses the truth value of x_i.)

- In rewriting the quadratic program into a vector program, we must be careful: A vector program may refer *only* to the scalar products of the unknown vectors.

  For example, we cannot set a vector to a constant. So we represent the constant 1 by an arbitrary unit vector $\mathbf{e}$. The constraint $y_i(1 - y_i) = 0$ then becomes $\mathbf{t}_i^T(\mathbf{e} - \mathbf{t}_i) = 0$ (or, written another way, $\mathbf{t}_i^T\mathbf{e} = \mathbf{t}_i^T\mathbf{t}_i$).

12.3.4 Here is the resulting semidefinite relaxation (which we call the *basic* semidefinite relaxation), shown on a (trivial) concrete example:

The basic semidefinite relaxation of MAX-2-SAT shown for the formula $(x_1 \vee x_2) \wedge (\overline{x}_2 \vee x_4) \wedge (x_1 \vee \overline{x}_3)$

$$
\begin{aligned}
\text{Maximize} \quad & \mathbf{t}_1^T\mathbf{t}_2 + \mathbf{t}_1^T(\mathbf{e} - \mathbf{t}_2) + (\mathbf{e} - \mathbf{t}_1)^T\mathbf{t}_2 \\
& + (\mathbf{e} - \mathbf{t}_2)^T\mathbf{t}_4 + (\mathbf{e} - \mathbf{t}_2)^T(\mathbf{e} - \mathbf{t}_4) + \mathbf{t}_2^T\mathbf{t}_4 \\
& + \mathbf{t}_1^T(\mathbf{e} - \mathbf{t}_3) + \mathbf{t}_1^T\mathbf{t}_3 + (\mathbf{e} - \mathbf{t}_1)^T(\mathbf{e} - \mathbf{t}_3) \\
\text{subject to} \quad & \mathbf{e}^T\mathbf{e} = 1 \\
& \mathbf{t}_i^T(\mathbf{e} - \mathbf{t}_i) = 0 \text{ for all } i.
\end{aligned}
$$

- For every other Boolean 2-CSP, we define the basic semidefinite relaxation using the same recipe. The only change compared to MAX-2-SAT is in the objective function, where we add terms corresponding to satisfying assignments of each predicate.

12.3.5 Here is the other "Goemans–Williamson" way of encoding the truth values.

- In the quadratic program, True is now represented by $+1$, and False by -1. The constraint $y_i^2 = 1$ restricts y_i to these two values.
- In the semidefinite relaxation, we replace y_i by a vector variable $\mathbf{v}_i$. Then $y_i^2 = 1$ translates to $\|\mathbf{v}_i\|^2 = 1$.
- As in the previous encoding with the $\mathbf{t}_i$, we also need a unit vector representing a constant (analogous to $\mathbf{e}$). Here it is traditionally denoted by $\mathbf{v}_0$. Intuitively, $\mathbf{v}_0^T\mathbf{v}_i = +1$ means $x_i = $ True and $\mathbf{v}_0^T\mathbf{v}_i = -1$ means $x_i = $ False.[1]
- This setting is perhaps more intuitive geometrically (we deal with unit vectors).
- It is easy to translate between these two settings, using the relations $\mathbf{e} = \mathbf{v}_0$ and $\mathbf{t}_i = \frac{1}{2}(\mathbf{v}_0 + \mathbf{v}_i)$.

---

[1] This meaning of $\mathbf{v}_0$ was introduced by Goemans and Williamson [GW95]. For increased confusion, more recent papers use $\mathbf{v}_0$ in opposite meaning, -1 or False. Then, of course, the resulting formulas again look different.

- For MAXCUT, $\mathbf{v}_0$ does not show up in the resulting objective function and, by this method, we get the usual semidefinite relaxation.
- For MAX-2-SAT, the contribution of the clause $x_1 \lor x_2$ now appears as $\frac{1}{4}(3 + \mathbf{v}_0^T \mathbf{v}_1 + \mathbf{v}_0^T \mathbf{v}_2 - \mathbf{v}_1^T \mathbf{v}_2)$ in the objective function.

12.3.6 For the basic semidefinite relaxation of MAX-2-SAT (in the version with the $\mathbf{v}_i$), we can apply the Goemans–Williamson random hyperplane rounding. Their analysis, almost without change, yields approximation ratio (at least) α_{GW}.

12.3.7 On the other hand, the integrality gap of the basic MAX-2-SAT relaxation is at least $\frac{9}{8} = 1.125$.

- This occurs for the instance with a single clause $x_1 \lor x_2$. We set $\mathbf{e} := (1,0)$, $\mathbf{t}_1 := \left(\frac{3}{4}, \frac{\sqrt{3}}{4}\right)$, $\mathbf{t}_2 := \left(\frac{3}{4}, -\frac{\sqrt{3}}{4}\right)$.

Then the constraints $\mathbf{t}_i^T (\mathbf{e} - \mathbf{t}_i) = 0$ hold (right?), but the clause is "over-satisfied": The objective function, $\mathbf{t}_1^T \mathbf{t}_2 + \mathbf{t}_1^T (\mathbf{e} - \mathbf{t}_2) + (\mathbf{e} - \mathbf{t}_1)^T \mathbf{t}_2$, attains value $9/8$.

12.3.8 It is thus natural to add constraints to the semidefinite program in order to reduce the integrality gap. For example, we can add constraints prohibiting "over-satisfaction" of clauses.

Adding Triangle Constraints

12.3.9 Adding constraints to a linear or semidefinite relaxation of an integer program is an old method for reducing the integrality gap. The additional constraints ("cutting planes") should be *valid*, i.e., preserve all integral solutions, but they should make the set of feasible solutions smaller.

- This can be done in a systematic way, for an arbitrary integer program, and organized in hierarchies of successively tighter relaxations (but with increasing number of constraints).
- There are (at least) three well-known hierarchies of this kind: the *Lovász–Schrijver hierarchy*, the *Sherali–Adams hierarchy*, and the **Lasserre hierarchy**. The first two concern both linear programming relaxations and semidefinite relaxations, the last one produces semidefinite relaxations.
- We will not treat them in this book, although they are conceptually important. We refer to the surveys by Laurent [Lau03] and by Chlamtac and Tulsiani [CT11].

- There have been recent *negative* results concerning these hierarchies. For example, for some NP-complete problems, it has been shown that even $\Omega(n)$ levels in these hierarchies are not sufficient to improve the approximation factor (where n is the number of vertices of an input graph, or some other measure of size of the input instance). See, e.g., Tulsiani [Tul09] and Raghavendra and Steurer [RS09a] and references therein.

12.3.10 In the previous section, we saw that a natural constraint to be added to the basic semidefinite relaxation of MAX-2-SAT is one bounding the contribution of each clause to the objective function by 1. For example, for the clause $x_1 \vee x_2$, such a constraint is

$$\mathbf{t}_1^T \mathbf{t}_2 + \mathbf{t}_1^T(\mathbf{e} - \mathbf{t}_2) + (\mathbf{e} - \mathbf{t}_1)^T \mathbf{t}_2 \leq 1.$$

- Using $\|\mathbf{e}\|^2 = 1$, it can be rewritten in the more elegant form

$$(\mathbf{e} - \mathbf{t}_1)^T(\mathbf{e} - \mathbf{t}_2) \geq 0.$$

- When we translate it to the $\mathbf{v}_i$-language ($\mathbf{v}_i = 2\mathbf{t}_i - \mathbf{e}$, $\mathbf{v}_0 = \mathbf{e}$), we obtain $(\mathbf{v}_0 - \mathbf{v}_1)^T(\mathbf{v}_0 - \mathbf{v}_2) \geq 0$. This tells us that in the triangle $\mathbf{v}_0\mathbf{v}_1\mathbf{v}_2$, the angle at the vertex $\mathbf{v}_0$ is at least $90°$.
- It is also equivalent to the following inequality for squared distances (using the cosine theorem):

$$\|\mathbf{v}_1 - \mathbf{v}_0\|^2 + \|\mathbf{v}_0 - \mathbf{v}_2\|^2 \leq \|\mathbf{v}_1 - \mathbf{v}_2\|^2.$$

This is a *triangle inequality* for the *squared* Euclidean distance on the set $\{\mathbf{v}_0, \mathbf{v}_1, \mathbf{v}_2\}$.

  ○ Note that the squared Euclidean distance on arbitrary three points need not satisfy the triangle inequality; for example, for three distinct collinear points, the triangle inequality always fails (one of the three triangle inequalities, that is)!

- For the other three possible forms of clauses ($x_1 \vee \overline{x}_2$, $\overline{x}_1 \vee x_2$, and $\overline{x}_1 \vee \overline{x}_2$) one obtains similar-looking (but different) constraints. Because of the above geometric meanings, they are called *triangle constraints*.

  ○ As we will see, they are the "best" constraints, in some sense, that one can add to the basic semidefinite relaxation of 2-CSP's.

12.3.11 Let us list the triangle constraints explicitly:

> Let us consider the basic semidefinite relaxation of a 2-CSP, with the unit vector $\mathbf{e}$ representing the constant 1 and with values of the variables represented by the vectors $\mathbf{t}_1, \ldots, \mathbf{t}_n$. Then by the *triangle constraints for* $\mathbf{e}$, $\mathbf{t}_i$, *and* $\mathbf{t}_j$ we mean the following (valid) constraints:
>
> $$\mathbf{t}_i^T \mathbf{t}_j \geq 0$$
> $$\mathbf{t}_i^T (\mathbf{e} - \mathbf{t}_j) \geq 0$$
> $$(\mathbf{e} - \mathbf{t}_i)^T \mathbf{t}_j \geq 0$$
> $$(\mathbf{e} - \mathbf{t}_i)^T (\mathbf{e} - \mathbf{t}_j) \geq 0.$$

The Power of the Triangle Constraints

12.3.12 We define the *canonical semidefinite relaxation* of a 2-CSP as the basic semidefinite relaxation plus the triangle constraints for $\mathbf{e}, \mathbf{t}_i, \mathbf{t}_j$, for all i, j such that x_i and x_j occur *together* in some clause of the input instance.

- Some justification for the word "canonical" will be provided later on.[2]
- Although some authors set up semidefinite relaxations of 2-CSP with triangle constrains for all pairs i, j, the analysis in these papers uses only those corresponding to variables in a common clause. Omitting the other triangle constraints makes the number of constraints linear in $m+n$. Moreover, it will be essential in a rounding algorithm in the next chapter.

12.3.13 Passing from the basic relaxation of MAX-2-SAT to the canonical one does indeed reduce the integrality gap.

- Lewin et al. [LLZ02], slightly improving on previous results by Feige and Goemans [FG95], proved that the integrality gap of the canonical semidefinite relaxation of MAX-2-SAT is at most 1.064 (while, as we saw, for the basic relaxation it is at least 1.125). They provided a rounding algorithm with the corresponding approximation ratio of at least $1/1.064 \approx 0.940$.

  ○ Although they state the result with all possible triangle constraints added, for all pairs i, j, their proof uses only triangle constraints of the canonical relaxation, i.e., involving variables occurring in a common clause.

---

[2] We borrowed this term from Karloff and Zwick [KZ97], who use it in the context of MAX-3-SAT, but in the 2-CSP context their definition would be equivalent to ours.

- The analysis is complicated and we will not discuss it.
- There is a quite close inapproximability bound of 0.954 for MAX-2-SAT (Håstad [Hås01], assuming only $P \neq NP$, as well as 0.943 under the UGC [KKMO07]).

12.3.14 For other 2-CSP, adding triangle inequalities does not necessarily help. For example, for MAXCUT, it can easily be shown that the optimal SDP solution for the Hamming graph considered in Sect. 8.4 satisfies all triangle constraints (and all other valid constraints, for that matter). Thus, the approximation ratio of the Goemans–Williamson algorithm (random hyperplane rounding) remains $1/\alpha_{GW}$ even if we add triangle constraints.

12.3.15 How about adding other kinds of constraints to the basic semidefinite relaxation – can it help, i.e., reduce the integrality gap? We do not know for sure. However:

- The UGC implies that for every problem MAX-2-CSP[$\mathcal{P}$], the canonical semidefinite relaxation with a suitable rounding algorithm yields the best (worst-case) approximation ratio achievable by any polynomial-time algorithm (Raghavendra [Rag08]; also see Austrin [Aus07]). We will not prove this, but later (Chap. 13) we will discuss the rounding algorithm.

  ○ Unfortunately, knowing that the approximation ratio is worst-case optimal (assuming the UGC) does not mean that we can actually determine it. For example, for MAX-2-SAT, assuming the UGC, only the bounds 0.940 from below and 0.943 from above mentioned earlier seem to be known.

12.3.16 The triangle constraints imply all other possible "local" valid constraints, as we now explain.

- Let us consider a clause in the input instance containing the variables x_i and x_j. We define a *local constraint* for that clause as a linear inequality of the form
$$a_1 \mathbf{e}^T \mathbf{t}_i + a_2 \mathbf{e}^T \mathbf{t}_j + b \mathbf{t}_i^T \mathbf{t}_j + c \geq 0,$$
for some real coefficients a_1, a_2, b, c. A local constraint *for a given instance* is one that is local for some of the clauses of the instance; in other words, if the corresponding two variables occur together in some of the clauses.
- Such a constraint is *valid* if it is satisfied by all integral solutions. That is, it holds for the 1-dimensional vectors $\mathbf{e} = 1$ and all choices of $\mathbf{t}_i, \mathbf{t}_j \in \{0, 1\}$.

12.3.17 Claim. *If $\tilde{\mathbf{e}}, \tilde{\mathbf{t}}_i, \tilde{\mathbf{t}}_j$ are vectors satisfying the triangle constraints (and otherwise completely arbitrary), then they also satisfy all valid constraints as in 12.3.16.*

12.3.18 Proof:

- Let us consider the set

$$S := \Big\{ \boldsymbol{\xi} \in \mathbb{R}^3 : \text{there exist vectors } \tilde{\mathbf{e}}, \tilde{\mathbf{t}}_i, \tilde{\mathbf{t}}_j$$

$$\text{satisfying the triangle constraints and}$$

$$\text{such that } \xi_1 = \tilde{\mathbf{e}}^T \tilde{\mathbf{t}}_i, \xi_2 = \tilde{\mathbf{e}}^T \tilde{\mathbf{t}}_j, \xi_3 = \tilde{\mathbf{t}}_i^T \tilde{\mathbf{t}}_j \Big\}.$$

- The triangle constraint $\mathbf{t}_i^T \mathbf{t}_j \geq 0$ implies that S is contained in the half-space $H_1 := \{ \boldsymbol{\xi} \in \mathbb{R}^3 : \xi_3 \geq 0 \}$. Similarly, the remaining three triangle constraints imply that S is also contained in the halfspaces H_2, H_3, H_4 with equations $\xi_1 - \xi_3 \geq 0$, $\xi_2 - \xi_3 \geq 0$, $1 - \xi_1 - \xi_2 + \xi_3 \geq 0$, respectively. The intersection of these halfspaces is the tetrahedron T shown below:

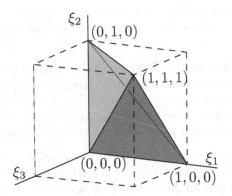

- Now let $a_1 \mathbf{e}^T \mathbf{t}_i + a_2 \mathbf{e}^T \mathbf{t}_j + b \mathbf{t}_i^T \mathbf{t}_j + c \geq 0$ be a valid local constraint, and let $H := \{ \boldsymbol{\xi} \in \mathbb{R}^3 : a_1 \xi_1 + a_2 \xi_2 + b \xi_3 + c \geq 0 \}$ be the corresponding halfspace in $\mathbb{R}^3$.
- Since the constraint is valid, it is satisfied for $\mathbf{e} = 1$ and all choices of $\mathbf{t}_i$ and $\mathbf{t}_j$ in $\{0, 1\}$, and so the points $(0, 0, 0)$, $(1, 0, 0)$, $(0, 1, 0)$, and $(1, 1, 1)$ lie in H. Then their convex hull also lies in H.
- But this convex hull is exactly the tetrahedron T, as is easy to check (this is the heart of the proof). Since $S \subseteq T$, all vectors $\tilde{\mathbf{e}}, \tilde{\mathbf{t}}_i, \tilde{\mathbf{t}}_j$ satisfying the triangle constraints also satisfy the considered valid constraint. This proves the claim. $\square$

12.3.19 Thus, the triangle constraints imply all valid local constraints. Of course, if the UGC fails, it might still be possible to improve the semidefinite relaxation by adding non-local constraints, or to set up a completely different and better semidefinite relaxation, but nobody has managed to do so. The canonical semidefinite relaxation remains the strongest available one for all 2-CSP.

12.3.20 The canonical semidefinite relaxation of a 2-CSP can be approximately solved in time almost linear in m, the number of clauses (Steurer [Ste10], building on a work by Arora and Kale [AK07]).

- Here "approximately" is meant in the following sense: Let $\varepsilon > 0$ be a parameter, and let φ be an input 2-CSP formula with n variables and $m \geq n$ clauses. Assuming that the value of an optimal solution of the canonical relaxation for φ is αm, then the algorithm computes a solution with value at least $(\alpha - \varepsilon)m$ that is feasible for the canonical relaxation of a formula φ', obtained from φ by deleting at most εm clauses.
- The running time is bounded by $O(m\varepsilon^{-C_1}(\log n)^{C_2})$ with some constants C_1, C_2.

12.4 Beyond Binary Boolean: Max-3-Sat & Co

12.4.1 Now we want a good semidefinite relaxation for constraint satisfaction problems with more than two variables per clause. We explain it on a concrete example of Max-3-Sat, but it will be apparent that the same method works for any CSP with two-valued (Boolean) variables.

12.4.2 So we consider a 3-Sat formula φ, say $(x_2 \vee \overline{x}_5 \vee x_7) \wedge \cdots \wedge (\overline{x}_3 \vee x_4 \vee \overline{x}_5)$ with variables $x_1, \ldots, x_n$ and with m clauses.

- We want to keep the features that worked well for the Max-2-Sat case, so we again use the vector variables $\mathbf{e}$ and $\mathbf{t}_1, \ldots, \mathbf{t}_n$ with the same meaning, as well as the basic constraints $\mathbf{e}^T \mathbf{e} = 1$ and $\mathbf{t}_i^T(\mathbf{e} - \mathbf{t}_i) = 0$.
- But a problem comes when we try to write down the objective function. In the binary case, for example, we could use the term $\mathbf{t}_2^T \mathbf{t}_5$ to represent "$x_2 = $ True and $x_5 = $ True." But how do we express "$x_2 = $ True and $x_5 = $ True and $x_7 = $ True"? It seems that we need a *cubic* term, but this does not fit in the SDP context.

12.4.3 Here is an ingenious solution. (It can be viewed as an instance of the general lift-and-project approach; see, e.g., Laurent [Lau03] – but we will not go into this.)

- For the first clause, $x_2 \vee \overline{x}_5 \vee x_7$, we introduce $2^3 = 8$ new auxiliary scalar variables $z_{1,\omega} \geq 0$, where $\omega \in \{\mathsf{F}, \mathsf{T}\}^3$ (we abbreviate $\mathsf{F} = $ False and $\mathsf{T} = $ True).
- In an integral solution, corresponding to an actual assignment, the $z_{1,\omega}$ should attain values 0 or 1, and, for example, $z_{1,\mathsf{TFT}}$ should be 1 for an assignment with $x_2 = $ True, $x_5 = $ False, $x_7 = $ True, while $z_{1,\mathsf{TFT}} = 0$ for all other assignments. So for an integral solution, exactly one among the $z_{1,\omega}$ is 1, the one corresponding to the values of x_2, x_5, x_7. In the semidefinite relaxation we thus add the constraint

$$\sum_{\omega \in \{\mathsf{F},\mathsf{T}\}^3} z_{1,\omega} = 1.$$

- Now we can conveniently write down the contribution of the first clause $x_2 \vee \overline{x}_5 \vee x_7$ to the objective function as

$$z_{1,\mathsf{FFF}} + z_{1,\mathsf{FFT}} + z_{1,\mathsf{FTT}} + z_{1,\mathsf{TFF}} + z_{1,\mathsf{TFT}} + z_{1,\mathsf{TTF}} + z_{1,\mathsf{TTT}}$$

  ($z_{1,\mathsf{FTF}}$ is missing).
- Similarly we introduce the $z_{\ell,\omega}$ for every $\ell = 1, 2, \ldots, m$.
- This may seem only to postpone the real problem. We can write $z_{1,\mathsf{TFT}}$ in the objective function, but how can we make sure that it really corresponds to $x_2 = \mathsf{True}$, $x_5 = \mathsf{False}$, $x_7 = \mathsf{True}$? So far there is no connection between the $z_{\ell,\omega}$ and the $\mathbf{t}_i$.
- The solution is, we do what we can and see how it works.

  ○ First, $\mathbf{e}^T \mathbf{t}_2$ should express the truth of x_2, and x_2 should be true exactly if one of $z_{1,\mathsf{TFF}}$, $z_{1,\mathsf{TFT}}$, $z_{1,\mathsf{TTF}}$, $z_{1,\mathsf{TTT}}$ is 1. So we add the constraint

  $$\sum_{\omega \in \{\mathsf{F},\mathsf{T}\}^3:\, \omega_1 = \mathsf{T}} z_{1,\omega} = \mathbf{e}^T \mathbf{t}_2,$$

  and similar constraints for x_5 and x_7.
  ○ An analogous constraint corresponding to $x_2 = \mathsf{False}$ happens to be redundant, since $\mathbf{e}^T(\mathbf{e} - \mathbf{t}_2) = 1 - \mathbf{e}^T \mathbf{t}_2$ and $\sum_\omega z_{1,\omega} = 1$. So we ignore it.

- Second, $\mathbf{t}_2^T \mathbf{t}_5$ should express the truth value of $x_2 \wedge x_5$, and so we add the constraint

$$\sum_{\omega \in \{\mathsf{F},\mathsf{T}\}^3:\, \omega_1 = \mathsf{T}, \omega_2 = \mathsf{T}} z_{1,\omega} = \mathbf{t}_2^T \mathbf{t}_5,$$

  as well as analogous constraints with $\mathbf{t}_2^T \mathbf{t}_7$ and $\mathbf{t}_5^T \mathbf{t}_7$. Similar constraints for $(\mathbf{e} - \mathbf{t}_2)^T \mathbf{t}_5$ etc. are again redundant.
- We do this for each clause, and we have set up the *canonical semidefinite relaxation* of MAX-3-SAT.
- The same works for any MAX-k-CSP$[\mathcal{P}]$ with Boolean variables. The predicates play a role only in the objective function, while the constraints do not depend on them.

12.4.4 We summarize the general definition more formally. Let the input instance of MAX-k-CSP$[\mathcal{P}]$ be, as earlier, $P_1(x_{i_{11}}, x_{i_{12}}, \ldots, x_{i_{1k}}) \wedge \cdots \wedge P_m(x_{i_{m1}}, x_{i_{m2}}, \ldots, x_{i_{mk}})$.

The canonical semidefinite relaxation
of a Boolean MAX-k-CSP$[\mathcal{P}]$

Vector variables: $\mathbf{e}, \mathbf{t}_1, \ldots, \mathbf{t}_n$.
Scalar variables: $z_{\ell,\omega}$, $\ell = 1, 2, \ldots, m$, $\omega \in \{\mathsf{F}, \mathsf{T}\}^k$.

$$\text{Maximize} \sum_{\ell=1}^{m} \sum_{\omega \in \{\mathsf{F},\mathsf{T}\}^k : P_\ell(\omega)=\mathsf{T}} z_{\ell,\omega}$$

subject to $\mathbf{e}^T \mathbf{e} = 1$

$$\mathbf{t}_i^T(\mathbf{e} - \mathbf{t}_i) = 0 \qquad 1 \le i \le n$$

$$z_{\ell,\omega} \ge 0 \qquad 1 \le \ell \le m, \ \omega \in \{\mathsf{F}, \mathsf{T}\}^k$$

$$\sum_\omega z_{\ell,\omega} = 1 \qquad 1 \le \ell \le m$$

$$\sum_{\omega:\,\omega_j=\mathsf{T}} z_{\ell,\omega} = \mathbf{e}^T \mathbf{t}_{i_{\ell j}} \quad 1 \le \ell \le m, \ 1 \le j \le k$$

$$\sum_{\omega:\,\omega_j=\omega_{j'}=\mathsf{T}} z_{\ell,\omega} = \mathbf{t}_{i_{\ell j}}^T \mathbf{t}_{i_{\ell j'}}$$

$$1 \le \ell \le m, \ 1 \le j < j' \le k.$$

Having displayed the somewhat frightening notation for the general case, we return to the case of MAX-3-SAT, leaving the obvious generalizations to the reader.

12.4.5 It may be useful to think about the $z_{\ell,\omega}$ in a probabilistic language.

- Let us consider our concrete MAX-3-SAT example, where the first clause was $x_2 \vee \overline{x}_5 \vee x_7$.
- The $z_{1,\omega}$ are nonnegative numbers adding up to 1, so they specify a probability distribution on $\{\mathsf{F}, \mathsf{T}\}^3$. Then the constraints tell us that $\mathrm{Prob}[\omega_1 = \mathsf{T}] = \mathbf{e}^T \mathbf{t}_2$, where ω is drawn at random from the distribution given by the $z_{1,\omega}$, etc., and also that $\mathrm{Prob}[\omega_1 = \mathsf{T} \text{ and } \omega_2 = \mathsf{T}] = \mathbf{t}_2^T \mathbf{t}_5$, etc.

The Canonical Relaxation Implies All Local Valid Constraints

12.4.6 As in the $k = 2$ case, the UGC implies that with an appropriate rounding algorithm, the just defined canonical semidefinite relaxation yields the best possible approximation ratio for any polynomial-time algorithm (Raghavendra [Rag08], Raghavendra and Steurer [RS09b]).

12.4.7 An attentive reader of the previous sections may ask, where are the triangle constraints, which were a crucial part of the canonical relaxation for the $k = 2$ case?

12.4.8 A nice feature of the above approach is that they are already "there" and we need not add them explicitly. More generally, all valid local constraints are already there!

- Generalizing the notion introduced earlier in an obvious way, we define a *local constraint* for a clause C_ℓ, whose variables are $x_{i_1}, x_{i_2}, x_{i_3}$ (so i_1, i_2, i_3 depend on ℓ), as an inequality of the form

$$a_1 \mathbf{e}^T \mathbf{t}_{i_1} + a_2 \mathbf{e}^T \mathbf{t}_{i_2} + a_3 \mathbf{e}^T \mathbf{t}_{i_3} + b_{12} \mathbf{t}_{i_1}^T \mathbf{t}_{i_2} + b_{13} \mathbf{t}_{i_1}^T \mathbf{t}_{i_3} + b_{23} \mathbf{t}_{i_2}^T \mathbf{t}_{i_3}$$

$$+ c + \sum_{\omega \in \{\mathsf{F},\mathsf{T}\}^3} d_\omega z_{\ell,\omega} \geq 0$$

with some real coefficients $a_1, a_2, a_3, b_{12}, b_{13}, b_{23}, c$, and d_ω, $\omega \in \{\mathsf{F}, \mathsf{T}\}^3$. A local constraint for an instance is again a constraint that is local for some of the clauses.

- A constraint as above is *valid* if it is satisfied for every integral solution induced by an assignment to the variables x_i. By this we mean that each x_i is set to either 0 or 1, and then the variables in the semidefinite relaxation are set as follows:
  - $\circ$ $\mathbf{e} := 1 \in \mathbb{R}^1$
  - $\circ$ $\mathbf{t}_i := x_i \in \mathbb{R}^1$
  - $\circ$ $z_{\ell,\omega} := 1$ if the assignment to the variables in the ℓ-th clause equals ω, and $z_{\ell,\omega} := 0$ otherwise

12.4.9 We have the following generalization of Claim 12.3.17:

Claim. *Every feasible solution*

$$\mathbf{e}, (\mathbf{t}_1, \ldots, \mathbf{t}_n), \left(z_{\ell,\omega} : \ell = 1, \ldots, m, \ \omega \in \{\mathsf{F},\mathsf{T}\}^3 \right)$$

of the canonical semidefinite relaxation of MAX-3-SAT *(or of some other* MAX-k-CSP$[\mathcal{P}]$*) satisfies every valid local constraint.*

12.4.10 Proof:

- This is actually simpler than for the analogous claim in the $k = 2$ case. We write it down for MAX-3-SAT but it should be clear that the argument works for every MAX-k-CSP$[\mathcal{P}]$.
- Let us consider the first clause, for example. Let the coordinates in $\mathbb{R}^8$ be indexed by $\omega \in \{\mathsf{F}, \mathsf{T}\}^3$ and let $S \subseteq \mathbb{R}^8$ be the set of all vectors $\mathbf{z}_1 = (z_{1,\omega} : \omega \in \{\mathsf{F}, \mathsf{T}\}^3)$ that can appear as a part of a feasible solution.
- Let $T \subseteq \mathbb{R}^8$ be the set defined by the constraints $z_{1,\omega} \geq 0$, $\omega \in \{\mathsf{F}, \mathsf{T}\}^3$, and $\sum_\omega z_{1,\omega} = 1$. Every feasible solution satisfies these constraints, and so $S \subseteq T$.

- Geometrically, T is the 7-dimensional regular simplex in $\mathbb{R}^8$, whose vertices are the vectors $\mathbf{e}_\omega$, $\omega \in \{\mathsf{F}, \mathsf{T}\}^3$, of the standard orthonormal basis in $\mathbb{R}^8$. The next picture illustrates this for dimension 3 instead of 8:

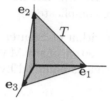

- Now let us consider some valid local constraint for the first clause. For every feasible solution of the canonical semidefinite relaxation, all the scalar products appearing in the constraint are linear combinations of the values of the $z_{1,\omega}$, and so the constraint becomes an inequality for the $z_{1,\omega}$:

$$c' + \sum_{\omega \in \{\mathsf{F}, \mathsf{T}\}^3} d'_\omega z_{1,\omega} \geq 0.$$

This defines a halfspace H in $\mathbb{R}^8$.
- We want to show that no feasible solution violates the constraint, i.e., that $S \subseteq H$. Since the constraint is valid, it is satisfied by all integral solutions derived from actual assignments to the variables x_i, as defined in 12.4.8. This implies that H contains all the vectors $\mathbf{e}_\omega$ as above, and hence also their convex hull T. But we know that $S \subseteq T$, and the claim is proved. $\quad\square$

Dealing with Multivalued Variables

12.4.11 So far we have considered constraint satisfaction problems with Boolean variables. Each such variable x_i was represented by a single vector $\mathbf{t}_i$ in the semidefinite relaxation.

12.4.12 If x_i attains values in a larger domain D, say $D = \{1, 2, \ldots, q\}$, a good solution is to introduce q vector variables $\mathbf{t}_{i,1}, \mathbf{t}_{i,2}, \ldots, \mathbf{t}_{i,q}$. The intended meaning is that $x_i = a$ should be represented by $\mathbf{e}^T \mathbf{t}_{i,a} = 1$ and $\mathbf{e}^T \mathbf{t}_{i,b} = 0$ for all $b \neq a$, $b \in D$.

12.4.13 We would thus like to add the constraint $\sum_{a \in D} \mathbf{t}_{i,a} = \mathbf{e}$ (for every i) to the semidefinite relaxation, but we cannot do it directly, since only scalar products may be used. However, we can formulate that constraint as $\left\| \mathbf{e} - \sum_{a \in D} \mathbf{t}_{i,a} \right\|^2 = 0$.

- Alternatively, we can replace $\mathbf{t}_{i,q}$ with $\mathbf{e} - \mathbf{t}_{i,1} - \cdots - \mathbf{t}_{i,q-1}$ everywhere. This is what we did in the case of $D = \{\mathsf{F}, \mathsf{T}\}$, where $\mathbf{t}_i$ played the role of $\mathbf{t}_{i,\mathsf{T}}$, while $\mathbf{t}_{i,\mathsf{F}}$ was represented implicitly as $\mathbf{e} - \mathbf{t}_i$.

12.4.14 Now it is straightforward to set up the canonical semidefinite relaxation of MAX-k-CSP$[\mathcal{P}]$ with an arbitrary domain D, imitating the Boolean case without much change, and check that the obvious generalization of Claim 12.4.9 goes through. We will not do it, since there is no new idea and the formalism gets more complicated.

• The near-linear time algorithm of Steurer [Ste10] mentioned in 12.3.20 works for the canonical relaxation of any MAX-k-CSP$[\mathcal{P}]$ with an arbitrary domain D. The running time is bounded by $O(m(k^{|D|}/\varepsilon)^{C_1}(\log n)^{C_2})$.

Exercises

12.1 Consider the following randomized algorithm for approximating MAX-3-SAT with exactly three distinct variables in each clause: Choose an assignment at random (tossing a fair coin for each variable), and repeat until at least $\lfloor \frac{7}{8}m \rfloor$ of the clauses are satisfied, where m is the total number of clauses. Prove that the expected running time is polynomial.

Hint: Use Markov's inequality for the number of *unsatisfied* clauses.

Chapter 13
Rounding Via Miniatures

13.1 An Ultimate Rounding Method?

13.1.1 The Goemans–Williamson semidefinite relaxation of MAXCUT is quite simple. Still, as we have seen, analyzing it, and proving tight bounds on the integrality gap, is not an easy task. For other basic problems, such as MAX-2-SAT or MAX-3-SAT, semidefinite relaxations proposed in the literature are more sophisticated.

For these semidefinite relaxations, more sophisticated rounding techniques have been proposed as well (pre-rounding rotations, skewed random hyperplane rounding, threshold rounding...). The analysis is complicated (sometimes computer-aided) and problem-specific.

13.1.2 We will discuss a recently discovered "generic" rounding technique, which attains the (worst-case) integrality gap, up to any given additive constant $\varepsilon > 0$, for a wide class of semidefinite relaxations of constraint satisfaction problems. ("A rounding to end all roundings...")

13.1.3 Moreover, assuming the Unique Games Conjecture, this rounding technique, applied to the canonical semidefinite relaxation introduced earlier, almost achieves the best approximation factor attainable by any polynomial-time algorithm, for every class of constraint satisfaction problems (Raghavendra [Rag08], Raghavendra and Steurer [RS09b]).

13.1.4 This is a wonderful result, but with some caveats:

- The algorithm is so far purely theoretical: the dependence on the constant parameters (ε, k, domain size) is doubly exponential. (A "revenge of asymptotic analysis.")
- Not clear what the worst-case approximation factor is. (The method also provides an algorithm for computing the approximation factor up to any given error ε, but that algorithm is doubly exponential and totally useless for an actual computation.)

B. Gärtner and J. Matoušek, *Approximation Algorithms and Semidefinite Programming*, DOI 10.1007/978-3-642-22015-9_13,
© Springer-Verlag Berlin Heidelberg 2012

- Does not apply to problems such as $\text{MaxQP}[G]$ (maximizing a quadratic form), where the objective function is a sum of both positive and negative contributions.
- Even for the "usual" constraint satisfaction problems, such as graph 3-coloring, we may be interested in a different kind of approximation guarantee: For example, we may want to control the fraction of *unsatisfied* constraints, rather than the satisfied ones. Or we want a proper coloring but allow for more than three colors. In such cases the method also has nothing to say at present.
- We should also remark that there are many classes of CSP where, under the UGC, the SDP machinery does not help at all. These are the so-called *approximation-resistant predicates*, for which the straightforward randomized algorithm, choosing the values of the variables independently and uniformly at random, achieves the best possible approximation ratio. See Austrin and Mossel [AM08].
- A strong example of approximation-resistance is a celebrated result of Håstad [Hås01] stating that it is NP-hard to approximate $\text{Max-3-CSP}[P]$ with ratio $\frac{1}{2} + \varepsilon$ for every fixed $\varepsilon > 0$, where the domain of the variables is $D = \{0, 1\}$ and $P(x_1, x_2, x_3)$ holds if $x_1 + x_2 + x_3 \equiv 1 \pmod 2$.

13.1.5 We will first show Raghavendra's rounding technique on the Goemans–Williamson semidefinite relaxation for MaxCut. There it brings no new result, but the basic idea can be illustrated in a simple form and in a familiar setting.

13.2 Miniatures for MaxCut

13.2.1 Let G be an input graph for MaxCut with vertex set $\{1, 2, \ldots, n\}$ as usual, and let $(\mathbf{v}_1, \ldots, \mathbf{v}_n)$ be an optimal SDP solution for it (i.e., the $\mathbf{v}_i$ are unit vectors maximizing $\sum_{\{i,j\} \in E} \frac{1 - \mathbf{v}_i^T \mathbf{v}_j}{2}$).

13.2.2 In the rounding algorithm, we build a *miniature* (= small model) of the input instance, which faithfully reflects the original, up to ε. (From now on, $\varepsilon > 0$ is arbitrarily small but fixed.)

- An important point: the miniature reflects not only the *input instance* G, but also the *optimal SDP solution* $(\mathbf{v}_1, \ldots, \mathbf{v}_n)$. It is a small graph $\hat{G}$ plus a sequence $(\hat{\mathbf{v}}_1, \ldots, \hat{\mathbf{v}}_{\hat{n}})$ of unit vectors.

13.2.3 A technical point: The graph in the miniature is *weighted*.

- By a *weighted graph* we mean a pair (G, w), where $G = (V, E)$ is a graph and $w \colon E \to (0, \infty)$ is a weight function on the edges.
- We write w_{ij} instead of $w(\{i, j\})$.
- Let $\text{Opt}(G, w)$ denote the maximum weight of a cut in (G, w), where the weight of a cut is the sum of the weights of its edges.

- The Goemans–Williamson algorithm clearly generalizes to such weighted graphs, with the same performance. The semidefinite relaxation is now

$$\mathsf{SDP}(G, w) := \max\left\{ \sum_{\{i,j\} \in E} w_{ij} \frac{1 - \mathbf{v}_i^T \mathbf{v}_j}{2} : \|\mathbf{v}_1\| = \cdots = \|\mathbf{v}_n\| = 1 \right\}.$$

- We let

$$\mathsf{Gap} := \sup_{G,w} \frac{\mathsf{SDP}(G, w)}{\mathsf{Opt}(G, w)}$$

be the worst-case integrality gap of the semidefinite relaxation, over all weighted graphs. (It is easily seen that Gap is the same as the maximum integrality gap for unweighted graphs. So we know from Sect. 1.3 and Theorem 8.3.2 that $\mathsf{Gap} = 1/\alpha_{\mathrm{GW}}$, but we will not use that.)

- We could also handle weighted *input* graphs without any additional difficulty; we stick to the unweighted case just to keep things (slightly) simpler.

13.2.4 So the miniature is a weighted $\hat{n}$-vertex graph $(\hat{G}, \hat{w})$ plus a sequence $\hat{\mathbf{v}}_1, \ldots, \hat{\mathbf{v}}_{\hat{n}}$ of unit vectors. We will make sure that the $\hat{\mathbf{v}}_i$ form a *feasible* solution (not necessarily optimal) of the semidefinite relaxation, as in 13.2.3, for $(\hat{G}, \hat{w})$, and that its value is almost the same as for the original SDP optimum, namely, at least $\mathsf{SDP}(G) - \varepsilon|E|$.

13.2.5 The miniature is so small that the maximum cut of $(\hat{G}, \hat{w})$ can be computed exactly by brute force. The number $\hat{n}$ of vertices is a (huge) constant, depending on ε.

The weighted graph $(\hat{G}, \hat{w})$ is a legal instance of the weighted MAXCUT problem, and so $\mathsf{Opt}(\hat{G}, \hat{w}) \geq \mathsf{SDP}(\hat{G}, \hat{w})/\mathsf{Gap} \geq (\mathsf{SDP}(G) - \varepsilon|E|)/\mathsf{Gap}$. (Here it is important that Gap was defined for weighted graphs, since even when we start with an unweighted graph, the miniature is weighted.)

13.2.6 From the optimal solution (maximum cut) for $(\hat{G}, \hat{w})$, we can reconstruct ("unfold") a cut in the original graph G with at least $\mathsf{Opt}(\hat{G}, \hat{w})$ edges.

- This is the output of the rounding algorithm. The number of edges of this cut is at least $(\mathsf{SDP}(G) - \varepsilon|E|)/\mathsf{Gap}$.
- This can be further bounded from below by $\mathsf{SDP}(G, w)(1 - 2\varepsilon)/\mathsf{Gap}$, since $\mathsf{SDP}(G) \geq \mathsf{Opt}(G) \geq \frac{1}{2}|E|$. (Indeed, for MAXCUT we know that the optimum has always at least half of the edges.)
- Thus, the approximation factor is at least $(1 - 2\varepsilon)/\mathsf{Gap}$.

13.2.7 A graphical summary: *Rounding via a miniature*

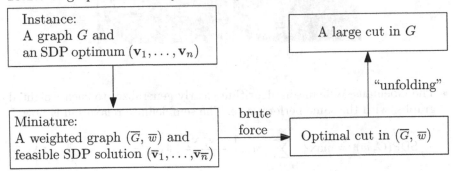

13.2.8 The technique of rounding via miniatures may remind some readers of applications of the *Szemerédi regularity lemma*. There are indeed some similarities, but a key difference is that the graph $\hat{G}$ in the miniature depends not only on the input graph, but also on the optimal SDP solution.

Two Lemmas for Miniature Builders

13.2.9 The first lemma tells us that unit vectors in S^{d-1} can be "discretized," up to a given error $\delta > 0$:

> **Lemma.** *For every d and every $\delta \in (0, 1)$, there exists a set $N \subseteq S^{d-1}$ that is δ-dense in S^{d-1} (that is, for every $\mathbf{x} \in S^{d-1}$ there exists $\mathbf{z} \in N$ with $\|\mathbf{x} - \mathbf{z}\| < \delta$), and $|N| \leq (\frac{3}{\delta})^d$.*

- This is a very useful and standard result, which one can hardly avoid in considerations involving high-dimensional geometry.
- We have seen the proof idea when discretizing the continuous graph in Sect. 8.3.

13.2.10 Proof:

- We build $N = \{\mathbf{p}_1, \mathbf{p}_2, \ldots\}$ by a "greedy algorithm": We place $\mathbf{p}_1$ to S^{d-1} arbitrarily, and having already chosen $\mathbf{p}_1, \ldots, \mathbf{p}_{i-1}$, we place $\mathbf{p}_i$ to S^{d-1} so that it has distance at least δ from $\mathbf{p}_1, \ldots, \mathbf{p}_{i-1}$.
- This process finishes as soon as we can no longer place the next point, i.e., the resulting set is δ-dense.
- To estimate the number of points produced in this way, we observe that the balls of radius $\frac{\delta}{2}$ centered at the $\mathbf{p}_i$ have disjoint interiors and are contained in the ball of radius $1 + \frac{\delta}{2} \leq \frac{3}{2}$ around $\mathbf{0}$. Thus, the sum of volumes of the small balls is at most the volume of the large ball, and this gives the lemma. $\qquad\square$

13.2.11 The second lemma is a special case of the *Johnson–Lindenstrauss lemma*, which says roughly the following: If $\mathbf{v}$ is a fixed vector in a "high-dimensional" Euclidean space $\mathbb{R}^n$ and $\varPhi\colon \mathbb{R}^n \to \mathbb{R}^d$ is a suitably chosen "normalized random linear map" into a "low-dimensional" space, then the image $\varPhi(\mathbf{v})$ has, with high probability, almost the same length as $\mathbf{v}$.

13.2.12 Since we will suffice with a weak quantitative bound for the "high probability," the proof of what we need is simpler than the "usual" proofs of the Johnson–Lindenstrauss lemma. On the other hand, we need an enhanced version asserting that a normalized random linear map also preserves, up to a small error, the scalar product of two vectors with high probability.

13.2.13 We use the following "normalized random linear map" $\varPhi\colon \mathbb{R}^n \to \mathbb{R}^d$. We choose d independent standard n-dimensional Gaussian vectors $\gamma_1, \ldots, \gamma_d \in \mathbb{R}^n$, and set $\varPhi(v) := \frac{1}{\sqrt{d}}(\gamma_1^T v, \ldots, \gamma_d^T v)$.

13.2.14 Lemma (Dimension reduction). *Let $\mathbf{u}, \mathbf{v} \in \mathbb{R}^n$ be unit vectors, let $\varPhi$ be the random linear map as above, and let $t \geq 0$. Then, for a sufficiently large constant C,*

$$\mathrm{Prob}\Big[|\mathbf{u}^T\mathbf{v} - \varPhi(\mathbf{u})^T\varPhi(\mathbf{v})| \geq t\Big] \leq \frac{C}{dt^2}.$$

13.2.15 Proof:

- Suffices to prove for $\mathbf{u} = \mathbf{v}$; the general case then follows using $\mathbf{u}^T\mathbf{v} = \frac{1}{2}(\|\mathbf{u}\|^2 + \|\mathbf{v}\|^2 - \|\mathbf{u} - \mathbf{v}\|^2)$. (The constant gets worse in this step, since we need to control the deviations for each of the three terms.)
- Let $X := \varPhi(\mathbf{v})^T\varPhi(\mathbf{v}) = \|\varPhi(\mathbf{v})\|^2$. Then $X = \frac{1}{d}\sum_{i=1}^d Z_i^2$, with $Z_i := \gamma_i^T\mathbf{v}$. The Z_i are standard normal (as we noted in 9.3.3) and independent, with $\mathrm{Var}[Z_i] = \mathbf{E}[Z_i^2] = 1$. So $\mathbf{E}[X] = 1$.
- The distribution of X is known as the *chi-square distribution*, and tail estimates for it can be found in the literature. For our purposes, there is also a simple direct proof.

- We use the *Chebyshev inequality*: $\mathrm{Prob}[\,|X - \mathbf{E}\,[X]| \geq t\,] \leq \mathrm{Var}\,[X]\,/t^2$, where $\mathrm{Var}\,[X] = \mathbf{E}\,[X^2] - \mathbf{E}\,[X]^2$ is the variance.
- By independence, $\mathrm{Var}\,[X] = \frac{1}{d^2}\sum_{i=1}^{d}\mathrm{Var}\,[Z_i^2]$.
- We want to see that each $\mathrm{Var}\,[Z_i^2]$ is bounded by a constant. There are several ways of checking this. A direct calculation (integration) gives $\mathrm{Var}\,[Z_i^2] = 2$. So $\mathrm{Var}\,[X] = \frac{2}{d}$ and the lemma follows. $\qquad\square$

Building the Miniature

13.2.16 We fix $\delta > 0$; this is the famous $\delta > 0$ which exists for every $\varepsilon > 0$.

13.2.17 We choose $\hat{d}$ sufficiently large in terms of δ. The vectors $\hat{\mathbf{v}}_i$ of the miniature will live in $\mathbb{R}^{\hat{d}}$.

13.2.18 We fix a δ-dense set $\hat{N}$ in $S^{\hat{d}-1}$ according to Lemma 13.2.9. Its size $\hat{n} := |\hat{N}|$ is going to be the number of vertices of the miniature graph $\hat{G}$.

13.2.19 We choose a random linear map $\Phi\colon \mathbb{R}^n \to \mathbb{R}^{\hat{d}}$, as in Lemma 13.2.14, and set $\mathbf{v}_i^* := \Phi(\mathbf{v}_i)$, $i = 1, 2, \ldots, n$.

13.2.20 We expect Φ to preserve most of the scalar products up to $\pm\delta$. Let us say that Φ *fails* on the vertex i if $\|\mathbf{v}_i^*\| \notin [1-\delta, 1+\delta]$. Similarly, Φ fails on an edge $\{i,j\}$ if it fails on i or j or if $\left|(\mathbf{v}_i^*)^T\mathbf{v}_j^* - \mathbf{v}_i^T\mathbf{v}_j\right| > \delta$.

13.2.21 Let $F \subseteq E$ be the set of edges where Φ fails. By Lemma 13.2.14 we have $\mathbf{E}\,[|F|] \leq \delta|E|$ (for $\hat{d}$ sufficiently large). So by Markov's inequality $|F| \leq 2\delta|E|$ with probability at least $\frac{1}{2}$. We repeat the random choice of Φ until $|F| \leq 2\delta|E|$ holds; the expected number of repetitions is at most 2.

13.2.22 We let G^* be the graph obtained from G by discarding all vertices and edges where Φ fails.

13.2.23 Next, we *discretize* the vectors $\mathbf{v}_i^*$: For every $i \in V(G^*)$, we choose $\mathbf{v}_i^{**} \in \hat{N}$ at distance at most 2δ from $\mathbf{v}_i^*$. (This is possible since each $\mathbf{v}_i^*$ is at most δ away from the sphere $S^{\hat{d}-1}$, and $\hat{N}$ is δ-dense in $S^{\hat{d}-1}$.)

13.2.24 Finally, we *fold* the graph G^* so that all vertices with the same vector $\mathbf{v}_i^{**}$ are identified into the same vertex. The miniature graph $(\hat{G}, \hat{w})$ is the result of this folding:

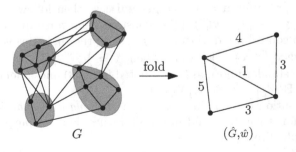

$G \qquad\qquad (\hat{G}, \hat{w})$

Explicitly:

- The vertices of $\hat{G}$ are $1, 2, \ldots, \hat{n}$.
- Let $\hat{\mathbf{v}}_1, \ldots, \hat{\mathbf{v}}_{\hat{n}}$ be the points (unit vectors) of $\hat{N}$ listed in some fixed order. These will form the feasible SDP solution for the miniature.
- The edges of $\hat{G}$ are obtained from the edges of G^*; i.e., $\{\hat{i}, \hat{j}\}$ is an edge if there is an edge $\{i, j\} \in E(G^*)$ such that $\mathbf{v}_i^{**} = \hat{\mathbf{v}}_{\hat{i}}$ and $\mathbf{v}_j^{**} = \hat{\mathbf{v}}_{\hat{j}}$.
- The weight of such an edge $\{\hat{i}, \hat{j}\}$ is the number of edges $\{i, j\}$ that got folded onto $\{\hat{i}, \hat{j}\}$:

$$\hat{w}_{\hat{i}\hat{j}} := \left| \left\{ \{i, j\} \in E(G^*) : \mathbf{v}_i^{**} = \hat{\mathbf{v}}_{\hat{i}}, \mathbf{v}_j^{**} = \hat{\mathbf{v}}_{\hat{j}} \right\} \right|.$$

13.2.25 Clearly, $\hat{\mathbf{v}}_1, \ldots, \hat{\mathbf{v}}_{\hat{n}}$ define a feasible SDP solution for $(\hat{G}, \hat{w})$ (they are unit vectors, and there are no other constraints to check – this is why MAXCUT is easy for the method of miniatures).

13.2.26 How much smaller can the SDP value $\sum_{\{\hat{i},\hat{j}\} \in E(\hat{G})} \hat{w}_{\hat{i}\hat{j}}(1 - \hat{\mathbf{v}}_{\hat{i}}^T \hat{\mathbf{v}}_{\hat{j}})/2$ be compared to $\mathsf{SDP}(G)$? Let us follow the algorithm backwards:

- The folding loses nothing; the value equals $\sum_{\{i,j\} \in E(G^*)} (1 - (\mathbf{v}_i^{**})^T \mathbf{v}_j^{**})/2$ (here we need to be slightly careful with vertices i, j that get folded to the same vertex $\hat{i}$ of $\hat{G}$, since there is no edge corresponding to $\{i, j\}$ in $\hat{G}$ – but for such i, j we have $(\mathbf{v}_i^{**})^T \mathbf{v}_j^{**} = 1$, and so they do not contribute to the SDP value).
- The discretization, replacing $\mathbf{v}_i^*$ by $\mathbf{v}_i^{**}$, changes each of the relevant scalar products $(\mathbf{v}_i^*)^T \mathbf{v}_j^*$ by at most $O(\delta)$ (additive error), and thus the objective function changes by no more than $O(\delta) \cdot |E|$.
- Similarly, for each non-failed edge $\{i, j\}$, replacing the original $\mathbf{v}_i$ and $\mathbf{v}_j$ by the Φ-images $\mathbf{v}_i^*, \mathbf{v}_j^*$ changes the scalar product $\mathbf{v}_i^T \mathbf{v}_j$ by at most δ, and again, the total change in the objective function is $O(\delta) \cdot |E|$.
- Finally, the number of failed edges is at most $2\delta|E|$, and discarding these may decrease the objective function by at most this much.
- Hence, $\mathsf{SDP}(\hat{G}, \hat{w}) \geq \mathsf{SDP}(G) - O(\delta) \cdot |E| \geq \mathsf{SDP}(G) - \varepsilon|E|$ as we wanted.

13.2.27 The miniature is finished. As announced earlier, its size is bounded by a constant, and we can compute $\mathsf{Opt}(\hat{G}, \hat{w})$ by brute force.

13.2.28 A cut in $(\hat{G}, \hat{w})$ of weight s can be *unfolded* to a cut in G^* with s edges, and hence to a cut with at least s edges in G.

13.2.29 This finishes the presentation of the rounding algorithm. We have shown the following:

> **Theorem.** *For every fixed $\varepsilon > 0$, and for every input graph G, the above rounding algorithm computes a cut of size at least*
>
> $$\frac{1}{(1+\varepsilon)\mathsf{Gap}} \cdot \mathsf{Opt}(G),$$
>
> *in expected polynomial time.*

- The dependence of the running time on ε is doubly exponential, unfortunately.
- The approximation factor is actually the reciprocal of the integrality gap for the miniature (up to ε). Even if the integrality gap for the input instance G is small, it can become large by the folding.
  - Of course, we cannot expect any polynomial-time algorithm to round close to the integrality gap for the input instance, since such an algorithm would compute an almost optimum cut, which is known to be NP-complete.

13.2.30 We actually get a stronger result.

- For $c \in [\frac{1}{2}, 1]$, let

$$\mathsf{Gap}(c) := \sup \left\{ \frac{\mathsf{SDP}(G, w)}{\mathsf{Opt}(G, w)} : \mathsf{SDP}(G, w) \geq cw(E) \right\}$$

be the integrality gap for graphs whose SDP optimum is at least a c-fraction of the total edge weight.
- The miniature in the algorithm satisfies $\mathsf{SDP}(\hat{G}, \hat{w}) \geq \mathsf{SDP}(G) - \varepsilon|E|$. Thus, for graphs with $\mathsf{SDP}(G) \geq c|E|$, the algorithm computes a cut within $(1+\varepsilon)\mathsf{Gap}(c - \varepsilon)$ of optimum.

13.2.31 Here is a summary of the rounding algorithm.

> **Rounding via miniatures for MaxCut**
> Input: G and an optimal SDP solution $\mathbf{v}_1, \ldots, \mathbf{v}_n$.
>
> 1. (Dimension reduction) Map the vectors $\mathbf{v}_i$ into $\mathbb{R}^d$ using the random Gaussian map Φ. Repeat until only at most a small fraction of the edges fail.
> 2. Discard the failed vertices and edges.
> 3. Discretize the remaining vectors to vectors of a fixed δ-dense set.
> 4. Fold the graph, lumping together vertices with the same discretized vector.
> 5. Compute a maximum cut in the resulting constant-size graph.
> 6. Unfold back to a cut in the original graph.

13.3 Rounding the Canonical Relaxation of Max-3-Sat and Other Boolean CSP

13.3.1 Here we will extend the rounding technique shown in the previous section to deal with more complicated semidefinite relaxations of constraint satisfaction problems.

13.3.2 We will demonstrate it for the canonical semidefinite relaxation of Max-3-Sat and other Boolean CSP.

- As in the case of MaxCut, the approximation ratio will be expressed in terms of the worst-case integrality gap for the *weighted* version, in which each clause C_ℓ has a nonnegative real weight w_ℓ. We aim at maximizing the sum of weights of the satisfied clauses.
- In the semidefinite relaxation, the only change is adding the w_ℓ as coefficients in the objective function. Thus, for a set $\mathcal{P}$ of k-ary predicates, we set

$$\mathsf{Gap}_{\mathcal{P}} := \sup_{\varphi, w} \frac{\mathsf{SDP}(\varphi, w)}{\mathsf{Opt}(\varphi, w)},$$

where φ is an instance of Max-k-CSP$[\mathcal{P}]$ as introduced earlier, w is an assignment of nonnegative real weights to the clauses of φ, $\mathsf{Opt}(\varphi, w)$ is the maximum possible weight of satisfied clauses, and $\mathsf{SDP}(\varphi, w)$ is the optimum of the canonical semidefinite relaxation.

- As in the previous section, we present the rounding algorithm for an *unweighted* input instance (although there is almost no difference in handling weighted inputs).

13.3.3 Theorem. *For every fixed $\varepsilon > 0$, k, and a set $\mathcal{P}$ of k-ary predicates over the domain $D = \{\mathsf{F}, \mathsf{T}\}$, there is a randomized algorithm that, for every input instance φ of Max-k-CSP$[\mathcal{P}]$, computes an assignment satisfying at least*

$$\frac{1}{(1 + \varepsilon)\mathsf{Gap}_{\mathcal{P}}} \cdot \mathsf{Opt}(\varphi)$$

of the clauses of φ, in expected polynomial time.

13.3.4 The generalization to larger domains D is not too difficult, but at some points in the proof one needs to work somewhat harder.

- We assume that each predicate in $\mathcal{P}$ has at least one satisfying assignment (totally unsatisfiable predicates may perhaps make sense as a punishment, but in the enlightened world of theoretical computer science we can ignore them). Then $\mathsf{Opt}(\varphi) \geq 2^{-k}m$ for every input instance φ with m clauses, since the expected number of clauses satisfied by a random assignment is at least $2^{-k}m$.

13.3.5 The *rounding algorithm* itself is a straightforward generalization of the one for MAXCUT shown in the previous section. The new part is in the analysis.

13.3.6 In the algorithm we work only with the vector variables $\mathbf{e}, \mathbf{t}_1, \ldots, \mathbf{t}_n$ and ignore the $z_{\ell,\omega}$. Their time comes in the analysis.

13.3.7 Here is the algorithm explicitly.

- We start with an input formula φ and an optimal solution $\mathbf{s} = (\mathbf{e}, \mathbf{t}_1, \ldots, \mathbf{t}_n;$ $z_{\ell,\omega} : \ell = 1, \ldots, m, \omega \in \{\mathsf{F}, \mathsf{T}\}^k)$ of the canonical semidefinite relaxation. The first step is a dimension reduction: We set $\mathbf{e}^* := \Phi(\mathbf{e})$ and $\mathbf{t}_i^* := \Phi(\mathbf{t}_i)$ for the random Gaussian map $\Phi \colon \mathbb{R}^{n+1} \to \mathbb{R}^{\hat{d}}$. (Here $\hat{d} = \hat{d}(\delta)$, with $\delta = \delta(\varepsilon) > 0$ sufficiently small.)

- Φ fails for a clause C_ℓ if at least one of the scalar products $\mathbf{e}^T\mathbf{e}$, $\mathbf{e}^T\mathbf{t}_i$, $\mathbf{t}_i^T\mathbf{t}_j$, where x_i, x_j are variables in C_ℓ, is changed by more than δ (additive error). By restarting if necessary, we make sure that at most a δ-fraction of the clauses fail (using Lemma 13.2.14). In particular, $\mathbf{e}$ is all-important and we always restart if $(\mathbf{e}^*)^T\mathbf{e}^* \notin [1 - \delta, 1 + \delta]$.

- We discard all the failed clauses from the input formula, obtaining a formula φ^*. The corresponding terms in the objective function and the corresponding constraints of the semidefinite program are discarded as well, and so are the variables occurring only in failed clauses.

  - Here it is important that the canonical semidefinite relaxation includes only *local* constraints, and thus the number of constraints is proportional to the number of clauses. Indeed, the random projection will typically spoil a fixed fraction of the constraints, and so we do not want to have many more constraints than clauses.

- Now comes the discretization. For MAXCUT, we dealt only with (approximately) unit vectors, while here we have vectors of length 1 *or smaller*.

  - Indeed, since $\|\mathbf{e}\| = 1$, we get $\mathbf{e}^T\mathbf{t}_i \leq \|\mathbf{t}_i\|$. The constraint $\mathbf{t}_i^T(\mathbf{e} - \mathbf{t}_i) = 0$ can be rewritten as $\mathbf{e}^T\mathbf{t}_i = \|\mathbf{t}_i\|^2$, and thus we have $\|\mathbf{t}_i\| \geq \|\mathbf{t}_i\|^2$. So $\|\mathbf{t}_i\| \leq 1$ follows.

So we choose N as a δ-dense set in the unit *ball*, rather than unit sphere. No problem here; the proof of Lemma 13.2.9 applies to this case as well, actually with exactly the same bound. We obtain the discretized vectors, $\mathbf{e}^{**}$ and the $\mathbf{t}_i^{**}$ (where the indices i of the discarded variables are not considered).

- Folding the formula, by identifying all variables x_i with the same discretized vector $\mathbf{t}_i^{**}$ to a single variable, is done in an obvious way (we identify clauses with the same ordered k-tuple of variables and the same predicate). Let $(\hat{\varphi}, \hat{w})$ be the folded instance (with weights).

- We find an optimal assignment for $(\hat{\varphi}, \hat{w})$ by brute force, we unfold it to an assignment for φ, and this is the output of the algorithm.

13.3.8 Where is the problem with the analysis? We need to bound from below $\mathsf{SDP}(\hat{\varphi}, \hat{w})$, the semidefinite optimum for the miniature. In the MAX-CUT case, the $\hat{v}_i$ formed a feasible solution, whose value was almost the SDP value of the input instance, but here we do not have a feasible solution for $(\hat{\varphi}, \hat{w})$.

- First, we have not really defined the values of the $z_{\ell,\omega}$ variables for the miniature. But, for each of the clauses $\hat{C}_{\hat{\ell}}$ of the miniature, we can just take over the values of the $z_{\ell,\omega}$ from one of the clauses C_ℓ that got folded over to $\hat{C}_{\hat{\ell}}$.
- More seriously, though, the equality constraints, such as $\mathbf{e}^T \mathbf{e} = 1$ or $\mathbf{t}_i^T(\mathbf{e} - \mathbf{t}_i) = 0$, typically do not hold for the vectors of the miniature – they all hold approximately, up to $O(\delta)$, but not exactly.

13.3.9 We thus need to show that an "almost feasible" solution of the canonical semidefinite relaxation can be *fixed* to a truly feasible solution, without much loss in the value of the objective function. We state this as a general result about the canonical relaxation (and so we will no longer use the "miniature" notation).

13.3.10 For convenience, let us recall the constraints of the canonical semidefinite relaxation and give them names. There are:

- The **et**-*constraints*: $\mathbf{e}^T \mathbf{e} = 1$ and $\mathbf{t}_i^T(\mathbf{e} - \mathbf{t}_i) = 0$.
- The **et**z-*constraints* relating the appropriate sums of the $z_{\ell,\omega}$ to the scalar products of $\mathbf{e}, \mathbf{t}_1, \ldots, \mathbf{t}_n$, namely,

$$\sum_{\omega:\,\omega_j = \mathsf{T}} z_{\ell,\omega} = \mathbf{e}^T \mathbf{t}_{i_{\ell j}} \quad \text{and} \quad \sum_{\omega:\,\omega_j = \omega_{j'} = \mathsf{T}} z_{\ell,\omega} = \mathbf{t}_{i_{\ell j}}^T \mathbf{t}_{i_{\ell j'}}.$$

- The *nonnegativity constraints* $z_{\ell,\omega} \geq 0$.
- The *z-sum constraints* $\sum_\omega z_{\ell,\omega} = 1$.

13.3.11 We call

$$\tilde{\mathbf{s}} = (\tilde{\mathbf{e}}, \tilde{\mathbf{t}}_1, \ldots, \tilde{\mathbf{t}}_n; \tilde{z}_{\ell,\omega} : \ell = 1, \ldots, m, \ \omega \in \{\mathsf{F}, \mathsf{T}\}^k)$$

a δ-*almost feasible solution* of the canonical semidefinite relaxation (for some input formula φ, possibly weighted) if it satisfies all **et**-constraints and **et**z-constraints up to δ (i.e., the left-hand and right-hand sides of each constraint differ by at most δ), and satisfies the nonnegativity constraints and the z-sum constraints exactly.[1]

13.3.12 In view of the above discussion, the following proposition is all that is needed to finish the proof of Theorem 13.3.3.

Proposition (Fixing an almost-feasible solution). *For every $\varepsilon > 0$ (and every k) there exists $\delta > 0$ with the following property. Let $\tilde{\mathbf{s}}$ be a δ-almost*

---

[1] We could also assume that the nonnegativity constraints are satisfied only up to δ and easily fix them as well. But since in the algorithm we have them satisfied exactly, why bother.

feasible solution of the canonical semidefinite relaxation for an input instance (φ, w), and let M be the value of the objective function for $\tilde{s}$. Then there is a feasible solution with value at least $M - \varepsilon W$, where W is the total weight of the clauses of φ.

- In the analysis of the rounding algorithm we need this proposition as an *existence* result. We never need to actually *compute* the feasible solution.

13.3.13 The fixing proceeds in several steps, which we first outline and then treat in detail.

1. First we fix $\tilde{e}$. Namely, we set $e' := \tilde{e}/\|\tilde{e}\|$; this is a unit vector, and all constraints still hold up to $O(\delta)$ after replacing $\tilde{e}$ by e'.
2. Next, we fix the et-constraints. We replace each $\tilde{t}_i$ by t'_i, with $\|\tilde{t}_i - t'_i\| = O(\delta)$ and with the et-constraint $(t'_i)^T(e' - t'_i) = 0$ satisfied exactly. (Note that here we can deal with one i at a time.) For $(e', t'_1, \ldots, t'_n; \tilde{z}_{\ell,\omega})$ the et$z$-constraints again hold up to $O(\delta)$.
3. Now we produce the new $z'_{\ell,\omega}$, for all ℓ and ω, with $|z'_{\ell,\omega} - \tilde{z}_{\ell,\omega}| = O(\delta)$, so that $s' := (e', t'_1, \ldots t'_n; z'_{\ell,\omega})$ satisfies the et$z$-constraints (as well as the et-constraints and the z-sum constraint) exactly. However, we may break the nonnegativity constraints at this step; the $z'_{\ell,\omega}$ may be slightly negative.
4. Finally, we will adjust all the components of the solution once again, each of them by a small amount, and obtain the desired feasible solution s''. We need to pull the $z'_{\ell,\omega}$ slightly "inwards" into the simplex defined by the nonnegativity constraints and the z-sum constraints, and this is achieved by perturbing the solution s' obtained so far using a particular feasible solution, the *centroid solution* $\bar{s}$.

13.3.14 We will now execute this program step by step. Step 1 is clear, but the others will take some work.

Fixing the et-Constraints

13.3.15 In Step 2 we have a unit vector e' and a vector $\tilde{t}_i$ with $|(e' - \tilde{t}_i)^T \tilde{t}_i| = O(\delta)$, and we want to show that there is a new vector t'_i close to $\tilde{t}_i$ and such that $(e' - t'_i)^T t'_i = 0$.

13.3.16 There are several possible proofs. A quick one is obtained if we translate to the v_i-language (introduced in 12.3.5), setting $\tilde{v}_i := 2\tilde{t}_i - e'$. Then, as we know or can easily check, the goal becomes making $\tilde{v}_i$ unit, and for that, there is an easy recipe: set $v'_i := \tilde{v}_i/\|\tilde{v}_i\|$. Then we translate back: $t'_i := \frac{1}{2}(e' + v'_i)$.

13.3.17 It remains to estimate $\|\tilde{t}_i - t_i\|$, which is routine.

- We calculate $\|\tilde{\mathbf{v}}_i\|^2 = \|2\tilde{\mathbf{t}}_i - \mathbf{e}'\|^2 = 1 - 4(\mathbf{e}' - \tilde{\mathbf{t}}_i)^T\tilde{\mathbf{t}}_i = 1 + O(\delta)$. Hence $\|\tilde{\mathbf{v}}_i\| = 1 + O(\delta)$ as well.
- Then

$$
\begin{aligned}
\|\tilde{\mathbf{t}}_i - \mathbf{t}'_i\| &= \frac{1}{2}\|\tilde{\mathbf{v}}_i - \mathbf{v}'_i\| = \frac{1}{2}\left\|\tilde{\mathbf{v}}_i - \frac{\tilde{\mathbf{v}}_i}{\|\tilde{\mathbf{v}}_i\|}\right\| \\
&= \frac{1}{2}\|\tilde{\mathbf{v}}_i\|\left(1 - \frac{1}{\|\tilde{\mathbf{v}}_i\|}\right) \\
&= \frac{1}{2}(1 + O(\delta))\left(1 - \frac{1}{1 + O(\delta)}\right) = O(\delta).
\end{aligned}
$$

Step 2 is finished.

Fixing the et$z$-Constraints

13.3.18 We describe Step 3, where the vector variables $\mathbf{e}', \mathbf{t}'_1, \ldots, \mathbf{t}'_n$ remain fixed, and the $\tilde{z}_{\ell,\omega}$ are changed by a small amount so that they satisfy the et$z$-constraints, while the z-sum constraints $\sum_\omega z_{\ell,\omega} = 1$ also remain satisfied. However, we *do not* insist on satisfying the nonnegativity constraints $z_{\ell,\omega} \geq 0$. This step is simple linear algebra.

13.3.19 It suffices to treat one clause at a time. We will omit the clause index ℓ in the notation. We write $\tilde{\mathbf{z}} = (\tilde{z}_\omega : \omega \in \{\mathsf{F}, \mathsf{T}\}^k)$, and similarly for $\mathbf{z}'$.

13.3.20 Let $i_1, \ldots, i_k$ be the indices of the variables in the considered clause, and let

$$
\mathbf{p} := \left(1, (\mathbf{e}')^T\mathbf{t}'_{i_1}, \ldots, (\mathbf{e}')^T\mathbf{t}'_{i_k}, (\mathbf{t}'_{i_1})^T\mathbf{t}'_{i_2}, (\mathbf{t}'_{i_1})^T\mathbf{t}'_{i_3}, \ldots, (\mathbf{t}'_{i_{k-1}})^T\mathbf{t}'_{i_k}\right) \in \mathbb{R}^K,
$$

$K := 1 + k + \binom{k}{2}$, be the vector of the right-hand sides of the constrains we want $\mathbf{z}'$ to satisfy. We consider the coordinates in $\mathbb{R}^K$ indexed by subsets $J \subseteq \{1, 2, \ldots, k\}$ of size at most 2; then we can write, e.g., $p_{\{2,3\}} = (\mathbf{t}'_{i_2})^T\mathbf{t}'_{i_3}$, $p_{\{3\}} = (\mathbf{e}')^T\mathbf{t}'_{i_3}$, and $p_\emptyset = 1$.

13.3.21 We can formulate the constraints for the desired vector $\mathbf{z}' \in \mathbb{R}^{2^k}$ compactly as follows:

$$
f(\mathbf{z}') = \mathbf{p},
$$

where $f: \mathbb{R}^{2^k} \to \mathbb{R}^K$ is the linear mapping given by

$$
f(\mathbf{z})_J := \sum_{\omega \in \{\mathsf{F},\mathsf{T}\}^k : \omega[J] = \mathsf{T}} z_\omega, \quad |J| \leq 2
$$

(here $\omega[J] = \mathsf{T}$ means that the components of ω whose indices lie in J should be T; in particular, for $J = \emptyset$ we sum over all ω).

13.3.22 The assumption on $\tilde{\mathbf{z}}$ can be written as $\|f(\tilde{\mathbf{z}}) - \mathbf{p}\| = O(\delta).$[2] We want a $\mathbf{z}' \in \mathbb{R}^{2^k}$ with $f(\mathbf{z}') = \mathbf{p}$ and $\|\mathbf{z}' - \tilde{\mathbf{z}}\| = O(\delta)$. Here is a schematic geometric picture of the situation, with $\mathbb{R}^{2^k}$ indicated as a rectangle and $\mathbb{R}^K$ shown as a segment:

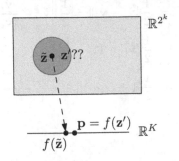

13.3.23 Since f is a linear map, it suffices to show that the vector $\mathbf{v} := \mathbf{p} - f(\tilde{\mathbf{z}})$ has a short preimage $\mathbf{u}$; that is, $f(\mathbf{u}) = \mathbf{v}$ and $\|\mathbf{u}\| = O(\|\mathbf{v}\|)$. Then we can set $\mathbf{z}' := \tilde{\mathbf{z}} + \mathbf{u}$ and we will be done.

13.3.24 Let A be the $K \times 2^k$ matrix of the linear mapping f (with respect to the standard bases). Its entries are $a_{J,\omega} = 1$ if $\omega[J] = \mathsf{T}$ and $a_{J,\omega} = 0$ otherwise. For example, for $k = 3$ we have $K = 7$ and

$$A = \begin{pmatrix} 1 & 1 & 1 & 1 & 1 & 1 & 1 & 1 \\ 0 & 0 & 0 & 0 & 1 & 1 & 1 & 1 \\ 0 & 0 & 1 & 1 & 0 & 0 & 1 & 1 \\ 0 & 1 & 0 & 1 & 0 & 1 & 0 & 1 \\ 0 & 0 & 0 & 0 & 0 & 0 & 1 & 1 \\ 0 & 0 & 0 & 0 & 0 & 1 & 0 & 1 \\ 0 & 0 & 0 & 1 & 0 & 0 & 0 & 1 \end{pmatrix}.$$

13.3.25 Let us suppose for a moment that A has rank K, which means that it has a nonsingular $K \times K$ submatrix B (for $k = 3$ it is easy to see and we will soon justify this assumption for all k). If needed, rearrange the columns of A so that B consists of the first K columns of A (for the A just shown no rearranging is needed).

13.3.26 Let us form the $2^k \times K$ matrix A^*, in which first K rows are the rows of B^{-1} and the remaining rows are $\mathbf{0}$'s. Then it is easily checked that $AA^* = I_k$ (so A^* is a *pseudoinverse* of A).

13.3.27 Now, with $\mathbf{v} = \mathbf{p} - f(\tilde{\mathbf{z}})$ as above, we set $\mathbf{u} := A^*\mathbf{v}$. We have $f(\mathbf{u}) = \mathbf{v}$ as desired, and it remains to bound $\|\mathbf{u}\|$.

---

[2] Strictly speaking, we should use the maximum norm and write $\|f(\tilde{\mathbf{z}}) - \mathbf{p}\|_\infty = O(\delta)$. But since the dimension K is a constant, the maximum norm and the Euclidean norm differ at most by a constant multiplicative factor, which can be swallowed by the $O(.)$ notation.

13.3.28 But since A^* depends only on k, and k is considered a constant, it is obvious that the linear map $\mathbf{y} \mapsto A^*\mathbf{y}$ can expand the length of any vector at most by a constant factor, and thus $\|\mathbf{u}\| = O(\|\mathbf{v}\|)$.

- We could easily come up with a quantitative bound of the form $2^{O(k)}$ on the expansion factor, by estimating the norm of the matrix B^{-1}.

13.3.29 Finally, why are the rows of A linearly independent? It is not hard to rearrange the rows and columns so that A has 1's on the main diagonal and 0's below it. We leave this as Exercise 13.1.

- Alternatively, one can extend A to a $2^k \times 2^k$ matrix $\overline{A}$, by letting J run through *all* subsets of $\{1, 2, \ldots, k\}$, and check that $\overline{A}$ is nonsingular. There are several ways of doing this; for example, one can write down an explicit inverse of $\overline{A}$.
- This $\overline{A}$ is an *inclusion matrix*; if we represent $\omega \in \{\mathsf{F}, \mathsf{T}\}$ by the set $I = \{j \in \{1, 2, \ldots, k\} : \omega_j = \mathsf{T}\}$, then $\overline{a}_{J,I} = 1$ for $J \subseteq I$ and $\overline{a}_{J,I} = 0$ otherwise. (It can also be related to the matrix of the discrete Fourier transform over the Boolean cube.)

Fixing Nonnegativity

13.3.30 In Step 4, we already have a solution $\mathbf{s}'$ satisfying all constraints except for the nonnegativity of the $z'_{\ell,\omega}$. We now define a particular feasible solution $\overline{\mathbf{s}}$, the *centroid solution*. The final feasible solution $\mathbf{s}''$ will be obtained as a "convex combination" of $\mathbf{s}'$ with $\overline{\mathbf{s}}$, in which $\overline{\mathbf{s}}$ has a small coefficient and $\mathbf{s}'$ has coefficient almost 1.

13.3.31 The centroid solution $\overline{\mathbf{s}}$ can be thought of as an average of all integral solutions, corresponding to all possible assignments to the variables x_i.

- The z-part of $\overline{\mathbf{s}}$ is thus defined by

$$\overline{z}_{\ell,\omega} := \frac{1}{2^k}$$

for all ω and ℓ.
- The $\mathbf{e}t z$-constraints are satisfied if $\overline{\mathbf{e}}^T \overline{\mathbf{t}}_i = \frac{1}{2}$ for all i and $\overline{\mathbf{t}}_i^T \overline{\mathbf{t}}_j = \frac{1}{4}$ for all i, j. This can be achieved by choosing $\overline{\mathbf{e}}$ and the $\overline{\mathbf{t}}_i$ in $\mathbb{R}^{2^k}$, with coordinates indexed by $\omega \in \{\mathsf{F}, \mathsf{T}\}$:

$$(\overline{\mathbf{e}})_\omega := 2^{-k/2},$$
$$(\overline{\mathbf{t}}_i)_\omega := \begin{cases} 2^{-k/2} & \text{if } \omega_i = \mathsf{T}, \\ 0 & \text{otherwise.} \end{cases}$$

It is easy to check that the $\overline{\mathbf{s}}$ thus defined is indeed feasible.

13.3.32 It remains to see how a "convex combination" of $\mathbf{s}'$ and $\bar{\mathbf{s}}$ is constructed.

- In the z-components, we will have a convex combination in the usual sense:

$$z''_{\ell,\omega} := (1 - \delta_1)z'_{\ell,\omega} + \delta_1 \bar{z}_{\ell,\omega},$$

  for a suitable $\delta_1 := C\delta$, C a sufficiently large constant.

- Since from the previous steps we have $z'_{\ell,\omega} \geq -C_1\delta$ for some constant C_1 (depending on k), for C large enough we get $z''_{\ell,\omega} \geq 0$ as desired. At the same time, the value of the objective function for $\mathbf{s}''$ differs from the value for the initial almost-feasible solution $\tilde{\mathbf{s}}$ by at most $O(\delta W)$, where W is the sum of weights of all clauses (since we have $\|z''_{\ell,\omega} - \tilde{z}_{\ell,\omega}\| = O(\delta)$ for all ℓ, ω).

- We still need to define a suitable operation for combining the vectors; we will use the *direct sum*. Let us assume that $\mathbf{e}', \mathbf{t}'_1, \ldots, \mathbf{t}'_n$ live in $\mathbb{R}^{d'}$ for some d', and let $\bar{\mathbf{e}}, \bar{\mathbf{t}}_1, \ldots, \bar{\mathbf{t}}_n$ live in $\mathbb{R}^{\bar{d}}$. Then we define $\mathbf{e}'', \mathbf{t}''_1, \ldots, \mathbf{t}''_n \in \mathbb{R}^{d'+\bar{d}}$ as

$$\mathbf{e}'' := \sqrt{1 - \delta_1}\mathbf{e}' \oplus \sqrt{\delta_1}\bar{\mathbf{e}},$$
$$\mathbf{t}''_i := \sqrt{1 - \delta_1}\mathbf{t}'_i \oplus \sqrt{\delta_1}\bar{\mathbf{t}}_i.$$

  That is, to get $\mathbf{e}''$, we write down the d' coordinates of $\mathbf{e}'$ multiplied by $\sqrt{1 - \delta_1}$, and then we append the $\bar{d}$ coordinates of $\bar{\mathbf{e}}$ multiplied by $\sqrt{\delta_1}$, and similarly for $\mathbf{t}''_i$.

- It is easy to check that this works. For arbitrary vectors $\mathbf{u}_1, \mathbf{u}_2, \mathbf{v}_1, \mathbf{v}_2$ we have $(\mathbf{u}_1 \oplus \mathbf{v}_1)^T(\mathbf{u}_2 \oplus \mathbf{v}_2) = \mathbf{u}_1^T\mathbf{v}_1 + \mathbf{u}_2^T\mathbf{v}_2$. It follows that the scalar products of $\mathbf{e}'', \mathbf{t}''_1, \ldots, \mathbf{t}''_n$ are convex combinations of the corresponding scalar products for $\mathbf{e}', \mathbf{t}'_1, \ldots, \mathbf{t}'_n$ and for $\bar{\mathbf{e}}, \bar{\mathbf{t}}_1, \ldots, \bar{\mathbf{t}}_n$, with coefficients $1 - \delta_1$ and δ_1. So $\mathbf{s}''$ is indeed a feasible solution of the canonical semidefinite relaxation.

  ○ The centroid solution $\bar{\mathbf{s}}$, which we have written down explicitly, can also be obtained from the integral solutions by the same operation as we used for combining $\mathbf{s}'$ and $\bar{\mathbf{s}}$ into $\mathbf{s}''$.

13.3.33 Proposition 13.3.12, as well as Theorem 13.3.3, are proved. $\qquad\square$

Exercises

13.1 (a) Let A be the matrix as in 13.3.24. Show that its rows are linearly independent, by rearranging to a triangular matrix.

(b) Let $\bar{A}$ be the inclusion matrix as in 13.3.29. Prove that it is nonsingular.

(c)* Let $1 \leq k < \ell \leq n/2$, and let A be the $\binom{n}{k} \times \binom{n}{\ell}$ inclusion matrix among all k-element subsets of $\{1, 2, \ldots, n\}$ and all ℓ-element subsets. Prove that A has the maximum rank, i.e., rank $\binom{n}{k}$.

Summary

Chapter 1

- A *c-approximation algorithm* for a maximization problem is a *polynomial-time* algorithm that computes, for every instance, a solution whose value is at least c times the optimal value.
- The running time is measured in the Turing machine model (and we need to be careful with arithmetic operations with large numbers!).
- Given a graph G, the MAXCUT problem asks for a set $S \subseteq V(G)$ such that the number of edges between S and its complement is maximum.
- The *Goemans–Williamson algorithm* is a randomized 0.878-approximation algorithm for MAXCUT, and it works as follows.
- We write the MAXCUT problem as a quadratic program, and we relax it to a vector program, looking for unit vectors $\mathbf{u}_1^*, \ldots, \mathbf{u}_n^*$ maximizing $\sum_{\{i,j\} \in E} (1 - \mathbf{u}_i^T \mathbf{u}_j)/2$. These can be computed (approximately) through semidefinite programming and Cholesky factorization.
- *Randomized rounding*: pick a random halfspace H through the origin and set $S := \{i : \mathbf{u}_i^* \in H\}$; this is the cut returned by the algorithm.
- In the analysis, for each edge $\{i, j\}$, we compare its contribution to the vector solution, which is $(1 - \mathbf{u}_i^{*T} \mathbf{u}_j^*)/2$, with the expected contribution to the resulting cut, which is $\arccos(\mathbf{u}_i^{*T} \mathbf{u}_j^*)/\pi$. The approximation factor $0.878\ldots$ is the minimum possible ratio of these quantities.

Chapter 2

- A *semidefinite program* generalizes a linear program. It is a problem of maximizing a linear function over the set of all positive semidefinite $n \times n$ matrices, subject to linear equality constraints involving the matrix entries.

B. Gärtner and J. Matoušek, *Approximation Algorithms and Semidefinite Programming*, DOI 10.1007/978-3-642-22015-9,
© Springer-Verlag Berlin Heidelberg 2012

- This is an *equational form* of a semidefinite program. More general semidefinite programs may also have inequality constraints and additional real (scalar) variables. They can easily be converted to equational form: for inequalities, we add new slack variables, and to represent a nonnegative real variable t, we enlarge the positive semidefinite matrix and put t to the diagonal, with 0's in the rest of the corresponding row and column.
- Unlike for linear programs, even if a semidefinite program has a finite value, the value need not be attained by any feasible solution.
- It is often said that semidefinite programs can be solved in polynomial time, but this has some caveats. The strongest theoretical result to date comes from the *ellipsoid method*, and it says that we can either find an ε-almost feasible and ε-almost optimal solution, or certify that the feasible region is ε-small. The running time is bounded by a polynomial in the input size *and* in $\log(R/\varepsilon)$, where R is some upper bound on the (Frobenius) norm of all feasible solutions. So a given semidefinite program can be approximately solved in polynomial time *provided* that we have an upper bound R with polynomially many digits.
- We exhibit an artificial semidefinite program for which the best R has exponentially many digits, meaning that we cannot always get polynomial runtime. However, in applications (such as MAXCUT), a good bound on R is usually obvious.
- For solving *vector programs*, we also need the *Cholesky factorization*, which decomposes a given positive semidefinite X as $X = U^T U$. We have presented a simple $O(n^3)$ algorithm.

Chapter 3

- The *strong product* $G \cdot H$ of graphs G and H has vertex set $V(G) \times V(H)$, and (u, u') and (v, v') are connected if u, v form an edge or coincide and also u', v' form an edge or coincide.
- The *Shannon capacity* of G is $\Theta(G) := \sup_k \alpha(G^k)^{1/k}$, where G^k is the k-fold strong product and $\alpha(.)$ is the independence number. This is motivated by coding theory, and it is one of the most tantalizing graph parameters.
- An *orthonormal representation* of a graph G is an assignment of unit vectors to vertices such that non-adjacent vertices receive orthogonal vectors. The *Lovász theta-function* $\vartheta(G)$ measures the smallest possible spherical cap containing all vectors of an orthonormal representation of G.
- We have $\Theta(G) \leq \vartheta(G)$ (the *Lovász bound*). The proof has two steps: first, $\alpha(G) \leq \vartheta(G)$, and second, $\vartheta(G \cdot H) \leq \vartheta(G)\vartheta(H)$ (via tensor product). As a (celebrated) consequence, we get $\Theta(C_5) = \vartheta(C_5) = \sqrt{5}$.
- $\vartheta(G)$ is (approximately) computable in polynomial time; a reformulation of the definition yields a semidefinite program.

- Another semidefinite program for $\vartheta(G)$ leads to an interpretation of $\vartheta(G)$ as the *vector chromatic number* of the complement $\overline{G}$. This yields the *sandwich theorem*: $\alpha(G) \leq \vartheta(G) \leq \chi(\overline{G})$. Consequently, for perfect graphs, $\alpha(G) = \chi(\overline{G})$ are computable in polynomial time.

Chapter 4

- For a semidefinite program $\max\{C \bullet X : A_1 \bullet X = b_1, \ldots, A_m \bullet X = b_m, X \succeq 0\}$, the *dual semidefinite program* is $\min\{\mathbf{b}^T\mathbf{y} : \sum_{i=1}^m y_i A_i - C \succeq 0\}$. The *strong duality theorem* asserts that if the primal SDP has a finite value β and some *positive definite* solution $\tilde{X}$, then the dual SDP also has value β.
- The proof is done in the more general framework of cone programming. Let V, W be Hilbert spaces (vector spaces with a scalar product), let $K \subseteq V$ and $L \subseteq W$ be closed convex cones, let $A: V \to W$ be a linear map, and let $\mathbf{c} \in V$, $\mathbf{b} \in W$. A *cone program* has the form $\max\{\langle \mathbf{c}, \mathbf{x} \rangle : \mathbf{x} \in K, \mathbf{b} - A(\mathbf{x}) \in L\}$. The dual cone program is $\min\{\langle \mathbf{b}, \mathbf{y} \rangle : A^T(\mathbf{y}) - \mathbf{c} \in K^*, \mathbf{y} \in L^*\}$, where the star denotes *dual cone* and A^T is the *adjoint* of A (in finite dimension, it corresponds to transposing the matrix of the linear map).
- For a semidefinite program in equational form, V is the space of all symmetric $n \times n$ matrices, $K = \mathrm{PSD}_n$ consists of all positive semidefinite matrices, $W = \mathbb{R}^m$, and $L = \{\mathbf{0}\}$. We have $\mathrm{PSD}_n^* = \mathrm{PSD}_n$.
- Cone programming has a nice duality theory, very similar to linear programming duality. The essential difference is that cone programs may exhibit *limit-feasibility*, meaning that an infeasible program may become feasible under an arbitrarily small perturbation of the constraints. Similarly, a cone program has a *limit value*, which may differ from the value.
- We start with a separation theorem for closed convex cones, and from this we prove a cone version of the Farkas lemma.
- The Farkas lemma is then used to prove the *regular duality theorem* of cone programming: the primal program is limit-feasible with limit value β if and only if the dual program is feasible with value β.
- Strong cone programming duality (primal value equals dual value) requires additional conditions, such as the existence of an interior feasible point (an $\tilde{\mathbf{x}} \in K$ with $\mathbf{b} - A(\tilde{\mathbf{x}}) \in \mathrm{int}(L)$, or, for the special case $L = \{\mathbf{0}\}$, $\tilde{\mathbf{x}} \in \mathrm{int}(K)$).

Chapter 5

- We present *Hazan's algorithm*, a recent and simple algorithm for approximately solving semidefinite programs. While the runtime of the ellipsoid method scales with $\log(1/\varepsilon)$, where ε is the approximation error, the runtime of Hazan's algorithm scales with $1/\varepsilon^3$ (this is the price to pay for the simplicity of the algorithm and its analysis).

- Hazan's algorithm works only for *trace-bounded* semidefinite programs (i.e., ones with an a priori upper bound on the trace of the matrix X). In practice, we can usually assume such a bound.

- In a first step, we show how every trace-bounded semidefinite program can be reduced through binary search to a feasibility problem with linear constraints over the *spectahedron* (the set of positive semidefinite matrices of trace 1).

- In a second step (and this is one of Hazan's main tricks), the feasibility problem is approximately solved by minimizing a convex penalty function (that penalizes constraint violations) over the spectahedron.

- For this task, the *Frank–Wolfe algorithm*, originally developed for quadratic programs in the 1950s, is used. In each iteration, we minimize a linear approximation of the convex function over the spectahedron (this amounts to *largest eigenvector* computations), and we make a carefully chosen step towards the minimizer.

- It can be shown that the error after k iterations is $O(1/k)$, where the constant depends on the curvature of the function. Computing the curvature for our penalty functions yields the runtime of Hazan's algorithm, in terms of the number of approximate largest eigenvector computations.

- To compute approximate eigenvectors, we use and analyze the *power method*, the simplest numerical method for this task. It is less efficient in the worst case than other methods, but if there is a single dominant eigenvalue, the method is very fast. The analysis for several dominant eigenvalues that we present is more involved but still elementary.

Chapter 6

- We sketch a *primal-dual central path algorithm* for approximately solving semidefinite programs in equational form, given that some initial pair $\tilde{X}, \tilde{y}$ of strictly feasible primal and dual solutions is available. This algorithm is from the family of *interior-point methods*.

- Similar to the ellipsoid method, the runtime scales with $\log(D/\varepsilon)$, where ε is the desired approximation error, and $D = \mathbf{b}^T \tilde{y} - C \bullet \tilde{X}$ is the *duality gap* of the given initial solution. But here the runtime bound holds only in the RAM model, and no explicit bounds are available on the size of the numbers appearing in the computation. The algorithm is simple and much more efficient in practice than the ellipsoid method.

- First, we define the *primal-dual central path* and show that it exists. For this, we add a *barrier term* $\mu \ln \det X$ to the objective function $C \bullet X$, where $\mu > 0$ is a parameter. Then we prove that, assuming strict primal and dual feasibility, this auxiliary problem has a unique optimal solution $X^*(\mu)$. The set of all $X^*(\mu), \mu > 0$ form the (primal part of the) central path. Moreover, for μ small, $X^*(\mu)$ is an almost optimal solution of the original semidefinite program.

- Then we express $X^*(\mu)$ as the zero of a multivariate (and nonlinear) function F_μ.

- The first idea is to compute the zero of F_μ by *Newton's method* in which each step boils down to a system of linear equations.

- But a problem with applying Newton's method directly is that it may lead us *outside* the set of positive definite matrices. (Additional technical complications are choosing F_μ so that the intermediate solutions stay symmetric, and also getting an initial solution sufficiently close to the central path.)

- The solution is path-following: assuming that we have a positive definite solution close to the central path at some (possibly large) value of μ, we make one Newton step w.r.t. a slightly smaller μ. If the parameters are carefully chosen, the step takes us to a positive definite solution that is close to the central path at the smaller μ. If we let μ decrease geometrically, we converge to a solution with error ε in time proportional to $\log(D/\varepsilon)$.

- One issue that we do not address in detail is how to get hold of the initial pair of solutions. The idea is to embed the semidefinite program together with its dual into a single larger program for which an initial solution is readily available. But this solution could have very large duality gap in the worst case, so that polynomial runtime is achieved only under additional conditions, similar to the ones in the ellipsoid method.

Chapter 7

- We introduce two closed convex cones. They are related to the cone of positive semidefinite matrices, but are hard to optimize over. A matrix M is called *copositive* if $\mathbf{x}^T M \mathbf{x} \geq 0$ for all $\mathbf{x} \geq \mathbf{0}$. The set of copositive matrices forms a cone COP_n containing the cone PSD_n of positive semidefinite matrices. The cone dual to COP_n is the cone POS_n of *completely positive* matrices which are defined as nonnegative linear combinations of rank-1 matrices. A *copositive program* looks like a semidefinite program, but with constraints of the form $X \in \mathrm{COP}_n$ or $X \in \mathrm{POS}_n$ instead of $X \succeq 0$.

- We relax the second semidefinite program for the Lovász theta function $\vartheta(G)$ introduced in Chap. 3, by replacing the constraint $X \succeq 0$ with $X \in$

COP$_n$. The value of the relaxation turns out to be $\alpha(G)$, the independence number of G. This already implies that copositive programming is hard.

- The proof uses the *Motzkin–Straus theorem*: the minimum value of the quadratic form $\mathbf{x}^T(A_G+I_n)\mathbf{x}$ over the unit simplex is $1/\alpha(G)$, where A_G is the adjacency matrix of G. As a corollary, we get that it is coNP-complete to decide whether a matrix is copositive.

- Using this, we prove that it is coNP-complete to decide whether a given point is a local minimizer of a smooth function over $\mathbb{R}^n$. (This shows that claims of reaching local optimality that are made by many numerical optimizers, have to be taken with caution.)

- The smooth functions that we use in the proof are of the form $(\mathbf{x}^2)^T M\mathbf{x}^2$, where $\mathbf{x}^2 = (x_1^2, x_2^2, \ldots, x_n^2)$. Then, $\mathbf{0}$ is a local minimizer of $(\mathbf{x}^2)^T M\mathbf{x}^2$ over $\mathbb{R}^n$ if and only if $\mathbf{0}$ is a local minimizer of $\mathbf{x}^T M\mathbf{x}$ over the positive orthant $\mathbb{R}_+^n$. The latter holds if and only if M is copositive.

Chapter 8

- In the Goemans–Williamson MAXCUT algorithm, and other algorithms based on a semidefinite relaxation, the *integrality gap* Gap is the largest ratio by which the maximum of the SDP can exceed the maximum of the original problem. Empirically, the approximation factor of such an algorithm is never better than $1/\mathsf{Gap}$.

- We construct graphs showing that $1/\mathsf{Gap}$ for the Goemans–Williamson algorithm equals $\alpha_{\mathrm{GW}} \approx 0.878\ldots$.

- First we take all points of a high-dimensional sphere S^d as vertices, and connect by edges the pairs with angle close to the worst-case angle ϑ_{GW}. Then we discretize this graph by subdividing S^d into tiny regions and picking a representative vertex for each region.

- In the proof we encounter *measure concentration*, in the form stating that most of the surface area of a high-dimensional spherical cap is concentrated near the boundary. We also meet *isoperimetric inequalities* and a proof method for them called *symmetrization*.

- Next, we construct graphs for which the Goemans–Williamson algorithm gives approximation no better than α_{GW}: the vertex set is $\{-1,1\}$, and the edges are pairs with a suitable Hamming distance corresponding to the angle ϑ_{GW}. In the proof we analyze the smallest eigenvalue of this graph, given by a binary Krawtchuk polynomial.

- Known: assuming the *Unique Games Conjecture* and $\mathrm{P} \neq \mathrm{NP}$, the approximation ratio α_{GW} is the best possible for any polynomial-time algorithm for MAXCUT.

- In the UGC, we consider systems of linear equations with two variables per equation, over a finite field of large but constant size, and the conjecture asserts that it is NP-hard to distinguish systems in which almost all equations are simultaneously satisfiable from those in which only a small fraction of the equations can be satisfied.

Chapter 9

- Given a 3-chromatic graph, we want to color it by as few colors as possible in polynomial time.
- An elementary trick (if there is a high-degree vertex, 2-color its neighborhood) yields $O(\sqrt{n})$ colors.
- An SDP-based algorithm guarantees roughly $n^{1/4}$ colors (current record is better). The key step is finding a large independent set.
- For this, take a *vector 3-coloring*, cut off a cap using a random halfspace generated via the d-dimensional *Gaussian distribution*, and take all isolated vertices within this cap.
- Difficult graphs for this algorithm have vector chromatic number 3 but chromatic number n^{δ}. They are obtained using theorems on forbidden intersections (e.g., take all $4t$-element subsets of $\{1, 2, \ldots, 8t\}$ as vertices and connect pairs with intersection size at most t).
- We present a proof with the linear-algebraic *polynomial method* and another using the *Vapnik–Chervonenkis–Sauer–Shelah lemma*.

Chapter 10

- The problems MaxCutGain, ground state in the Ising model, correlation clustering, and finding the *cut norm* of a matrix all lead to *maximizing a quadratic form on a graph*, i.e., $\max \sum_{\{i,j\} \in E} a_{ij} x_i x_j$, with $x_i \in [-1, 1]$.
- The *Grothendieck constant* K_G of a graph G is the maximum integrality gap of a natural SDP relaxation of this problem (with $x_i \in [-1, 1]$ replaced with $\mathbf{v}_i \in \mathbb{R}^n$, $\|\mathbf{v}_i\| \leq 1$).
- The main theorem claims that, for G loopless, $K_G = O(\log \vartheta(G))$. In particular, K_G is bounded by a constant for graphs of bounded chromatic number.
- The proof is an approximation algorithm. It rounds the SDP solution by taking scalar products of the vectors with a random Gaussian vector, truncating them to a suitable interval $[-M, M]$, and scaling down to $[-1, 1]$.
- Estimating the error made by the truncation is based on an interesting connection: the SDP optimum is an upper bound on the maximum *expectation* of the quadratic form, where we substitute random variables X_i with $\mathbf{E}\left[X_i^2\right] \leq 1$ for the x_i.

Chapter 11

- We consider a set system $\mathcal{F}$ on a finite ground set X, and want to color each point by $+1$ or -1 so that the *discrepancy* is small; the discrepancy is the maximum imbalance of the coloring over all $F \in \mathcal{F}$.
- Discrepancy is hard to approximate. We present an SDP-based algorithm that computes a good coloring for systems with small *hereditary discrepancy* (i.e., every restriction of $\mathcal{F}$ to a smaller ground set has a small discrepancy). This gives the first constructive version of several existential results for discrepancy, e.g., for arithmetic progressions.
- As an SDP relaxation of discrepancy we use the *vector discrepancy*: we color points with unit vectors and measure the discrepancy for a set as the length of the sum of the vectors of its points.
- The algorithm produces a sequence of *semicolorings*, in which the colors are reals in $[-1, 1]$. This is a random walk in the unit cube starting at $\mathbf{0}$ and proceeding in small increments until it reaches a vertex of the cube, i.e., a coloring.
- In each step of the walk, we want that the current discrepancies of all sets do not change too much. To this end, the direction of the step is generated from an auxiliary vector coloring via a random Gaussian projection.

Chapter 12

- We consider *constraint satisfaction problems* and their maximization versions.
- In MAX-3-SAT, we have Boolean variables $x_1, \ldots, x_n$ and m clauses of the form $x_i \vee \overline{x_j} \vee \overline{x_k}$ (some variables negated, some not), and want to satisfy as many clauses as possible. Here the best approximation ratio is known to be $\frac{7}{8}$.
- In a MAX-k-CSP$[\mathcal{P}]$, the variables can attain q values and each clause is a k-ary predicate from a given set $\mathcal{P}$ (k, q, $\mathcal{P}$ are considered fixed).
- The UGC in a more general (but equivalent) form deals with computational problems of the form MAX-2-CSP$[\mathcal{P}]$, for predicates corresponding to permutations and q large. Again it asserts NP-hardness of distinguishing instances with almost all clauses satisfiable from those with only a small fraction satisfiable.
- Boolean MAX-2-CSP's can be relaxed semidefinitely in a "canonical" way. The variables in the relaxation are vectors $\mathbf{t}_1, \ldots, \mathbf{t}_n$ of length at most 1, and an auxiliary unit vector $\mathbf{e}$, where $\langle \mathbf{e}, \mathbf{t}_i \rangle$ expresses the "degree of truth" of x_i. The clauses of the MAX-2-CSP are reflected in the objective function; e.g., the 2-SAT clause $x_2 \vee \overline{x}_4$ contributes $\langle \mathbf{t}_2, \mathbf{e} - \mathbf{t}_4 \rangle + \langle \mathbf{t}_2, \mathbf{t}_4 \rangle + \langle \mathbf{e} - \mathbf{t}_2, \mathbf{e} - \mathbf{t}_4 \rangle$ (one term for each satisfying assignment).

- To improve the integrality gap, one can add *triangle constraints*, essentially saying that no clause may contribute by more than one to the objective function. This yields the *canonical SDP relaxation*.
- Assuming the UGC, the canonical SDP relaxation with a suitable rounding leads to the best possible approximation ratio. The canonical relaxation can be solved in near-linear time.
- We show that the triangle constraints imply all valid local constraints.
- There are a similar canonical relaxation and similar results for q-valued MAX-k-CSP. Here the canonical SDP relaxation has variables $\mathbf{t}_{i,b}$, $i = 1, 2, \ldots, n$, $b = 1, 2, \ldots, q$, where $\langle \mathbf{e}, \mathbf{t}_{i,b} \rangle$ should reflect the truth values of $x_i = b$, and scalar variables $z_{\ell,\omega} \in [0, 1]$, where ℓ is a clause index and ω is a possible assignment to the variables in the clause. The purpose of $z_{\ell,\omega}$ is to capture the truth value of "the assignment to the variables of the ℓ-th clause equals ω."

Chapter 13

- The canonical relaxation of a MAX-k-CSP as in the previous chapter can be rounded in randomized polynomial time, with approximation factor no worse than $1/\text{Gap}$, where Gap is the maximum integrality gap for the considered class of MAX-k-CSP's. Assuming the UGC, this is the best approximation possible. The algorithm is not practical because of astronomically large constants.
- The rounding algorithm makes a projection of the optimal SDP solution to a random subspace of large constant dimension. From this, it builds a *miniature*, a constant-size MAX-k-CSP of the same type as the original problem.
- The miniature is solved by brute force, and the solution "unfolds" to a solution of the original MAX-k-CSP.
- The construction of the miniature relies on discretization using an ε-net, and on the *Johnson–Lindenstrauss lemma*, which asserts that scalar products of vectors are approximately preserved by a random projection with high probability.
- This works reasonably easily for MAXCUT. For other CSP's, we still need to show that an almost feasible solution of the SDP can be fixed to a truly feasible solution, with only a small loss in the objective function. This involves some interesting geometry.

References

[ABC97]   J. R. Alexander, J. Beck, and W. W. L. Chen. Geometric discrepancy
           theory and uniform distribution. In J. E. Goodman and J. O'Rourke,
           editors, *Handbook of Discrete and Computational Geometry*, pages 185–
           207. CRC Press LLC, Boca Raton, FL, 1997.

[ABH+05]   S. Arora, E. Berger, E. Hazan, G. Kindler, and M. Safra. On non-
           approximability for quadratic programs. In *FOCS'05: Proc. 46th IEEE
           Symp. on Foundations of Computer Science*, pages 206–215, 2005.

[ABS91]   N. Alon, L. Babai, and H. Suzuki. Multilinear polynomials and Frankl–
           Ray-Choudhuri Wilson type intersection theorems. *J. Comb. Theory,
           Ser. A*, 58(2):165–180, 1991.

[ABS10]   S. Arora, B. Barak, and D. Steurer. Subexponential algorithms for
           unique games and related problems. In *FOCS'10: Proc. 51st IEEE
           Symposium on Foundations of Computer Science*, pages 563–572, 2010.

[ACC06]   S. Arora, E. Chlamtac, and M. Charikar. New approximation guarantee
           for chromatic number. In *STOC'06: Proc. 38th ACM Symposium on
           Theory of Computing*, pages 215–224, 2006.

[ACOH+10] N. Alon, A. Coja-Oghlan, H. Hàn, M. Kang, V. Rödl, and M. Schacht.
           Quasi-randomness and algorithmic regularity for graphs with general
           degree distributions. *SIAM J. Comput.*, 39(6):2336–2362, 2010.

[ADRY94]   N. Alon, R. A. Duke, V. Rödl, and R. Yuster. The algorithmic aspects
           of the regularity lemma. *J. Algorithms*, 16:80–109, 1994.

[AHK05]   S. Arora, E. Hazan, and S. Kale. The multiplicative weights update
           method: a meta algorithm and applications, 2005. To appear in *Theory
           of Computing*.

[AK07]   S. Arora and S. Kale. A combinatorial, primal-dual approach to solv-
           ing SDPs. In *STOC'07: Proc. 39th ACM Symposium on Theory of
           Computing*, pages 227–236, 2007.

[AKK95]   S. Arora, D. Karger, and M. Karpinski. Polynomial time approximation
           schemes for dense instances of NP-hard problems. In *STOC'95: Proc.
           27th Annual ACM Sympos. on Theory of Computing*, pages 284–293,
           1995.

[AL11]   M. F. Anjos and J. B. Lasserre, editors. *Handbook on Semidefinite,
           Conic and Polynomial Optimization*. Springer, Berlin, 2012. To appear.

[Ali95]   F. Alizadeh. Interior point methods in semidefinite programming with
           applications to combinatorial optimization. *SIAM J. Opt.*, 5(1):13–51,
           1995.

[Alo98]   N. Alon. The Shannon capacity of a union. *Combinatorica*, 18:301–310,
           1998.

B. Gärtner and J. Matoušek, *Approximation Algorithms and Semidefinite   239
Programming*, DOI 10.1007/978-3-642-22015-9,
© Springer-Verlag Berlin Heidelberg 2012

[AM08]   P. Austrin and E. Mossel. Approximation resistant predicates from pair-
            wise independence. In *CCC'08: Proc. 23rd Annual IEEE Conference
            on Computational Complexity, Los Alamitos, CA*, pages 249–258. IEEE
            Computer Society, 2008.

[AMMN06]   N. Alon, K. Makarychev, Yu. Makarychev, and A. Naor. Quadratic
            forms on graphs. *Invent. Math.*, 163(3):499–522, 2006.

[AN96]   N. Alon and A. Naor. Approximating the cut-norm via Grothendieck's
            inequality. *SIAM J. Comput.*, 35(4):787–803, 1996.

[AS00]   N. Alon and B. Sudakov. Bipartite subgraphs and the smallest eigen-
            value. *Combin. Probab. Comput.*, 9(1):1–12, 2000.

[AS08]   N. Alon and J. Spencer. *The Probabilistic Method (3rd edition)*. J. Wiley
            and Sons, New York, NY, 2008.

[Aus07]   P. Austrin. Towards sharp inapproximability for any 2-CSP. In
            *FOCS'07: Proc. 48th IEEE Symp. on Foundations of Computer Sci-
            ence*, pages 307–317, 2007.

[Ban10]   N. Bansal. Constructive algorithms for discrepancy minimization.
            http://arxiv.org/abs/1002.2259, also in *FOCS'10: Proc. 51st IEEE
            Symposium on Foundations of Computer Science*, pages 3–10, 2010.

[BDDK+00]   I. M. Bomze, M. Dür, E. De Klerk, C. Roos, A. J. Quist, and T. Ter-
            laky. On copositive programming and standard quadratic optimization
            problems. *Journal of Global Optimization*, 18(4):301–320, 2000.

[Blu94]   A. Blum. New approximation algorithms for graph coloring. *J. Assoc.
            Comput. Mach.*, 41(3):470–516, 1994.

[BTN01]   A. Ben-Tal and A. Nemirovski. *Lectures on modern convex optimiza-
            tion. Analysis, algorithms, and engineering applications*. MPS-SIAM
            Series on Optimization. Society for Industrial and Applied Mathemat-
            ics (SIAM), Philadelphia, PA, USA, 2001.

[Bul02]   A. A. Bulatov. A dichotomy theorem for constraints on a three-element
            set. In *FOCS'02: Proc. 43rd IEEE Symp. on Foundations of Computer
            Science*, pages 649–658, 2002.

[BV04]   S. Boyd and L. Vandenberghe. *Convex Optimization*. Cambridge Uni-
            versity Press, Cambridge, 2004.

[Chl07]   E. Chlamtac. Approximation algorithms using hierarchies of semidefi-
            nite programming relaxations. In *FOCS'07: Proc. 48th IEEE Sympo-
            sium on Foundations of Computer Science*, pages 691–701, 2007.

[Cla10]   K. L. Clarkson. Coresets, sparse greedy approximation and the Frank-
            Wolfe algorithm. *ACM Transactions on Algorithms*, 6(4):63:1–63:30,
            2010.

[CNN11]   M. Charikar, A. Newman, and A. Nikolov. Tight hardness results for
            minimizing discrepancy. In *Proc. 22nd Annual ACM-SIAM Symposium
            on Discrete Algorithms (SODA)*, pages 1607–1614, 2011.

[CRST06]   M. Chudnovsky, N. Robertson, P. D. Seymour, and R. Thomas. The
            strong perfect graph theorem. *Ann. Math.*, 164:51–229, 2006.

[CT11]   E. Chlamtac and M. Tulsiani. Convex relaxations and integrality gaps.
            In M. F. Anjos and J. B. Lasserre, editors, *Handbook on Semidefinite,
            Conic and Polynomial Optimization*. Springer, Berlin, 2012. To appear;
            a preliminary version available from the authors' home pages.

[CW04]   M. Charikar and A. Wirth. Maximizing quadratic programs: extending
            Grothendieck's inequality. In *FOCS'04: Proc. 45th IEEE Symp. on
            Foundations of Computer Science*, pages 54–60, 2004.

[DMR06]   I. Dinur, E. Mossel, and O. Regev. Conditional hardness for approxi-
            mate coloring. In *STOC'06: Proc. 38th ACM Symposium on Theory of
            Computing*, pages 344–353, 2006.

[DR10]   I. Dukanovic and F. Rendl. Copositive programming motivated bounds on the stability and the chromatic numbers. *Math. Programming*, 121:249–268, 2010.

[Duf56]   R. J. Duffin. Infinite programs. In H. W. Kuhn and A. W. Tucker, editors, *Linear Inequalities and Related Systems*, volume 38 of *Annals of Mathematical Studies*, pages 157–170. 1956.

[EMT04]   Y. Eidelman, V. Milman, and A. Tsolomitis. *Functional Analysis: An Introduction*. Graduate Studies in Mathematics (Vol. 66). American Mathematical Society, 2004.

[FG95]   U. Feige and M. X. Goemans. Approximating the value of two prover proof systems, with applications to MAX 2-SAT and MAX DICUT. In *Proc. 3rd Israeli Symposium on Theory of Computing and Systems*, pages 182–189, 1995.

[FK98]   U. Feige and J. Kilian. Zero knowledge and the chromatic number. *J. Comput. Syst. Sci.*, 57(2):187–199, 1998.

[FK99]   A. Frieze and R. Kannan. Quick approximation to matrices and applications. *Combinatorica*, 19(2):175–220, 1999.

[FKL02]   U. Feige, M. Karpinski, and M. Langberg. Improved approximation of Max-Cut on graphs of bounded degree. *J. Algorithms*, 43:201–219, 2002.

[FL06]   U. Feige and M. Langberg. The RPR^2 rounding technique for semidefinite programs. *J. Algorithms*, 60(1):1–23, 2006.

[FS02]   U. Feige and G. Schechtman. On the optimality of the random hyperplane rounding technique for MAX CUT. *Random Struct. Algorithms*, 20(3):403–440, 2002.

[FV98]   T. Feder and M. Y. Vardi. The computational structure of monotone monadic SNP and constraint satisfaction: A study through Datalog and group theory. *SIAM J. Comput.*, 28(1):57–104, 1998.

[FW56]   M. Frank and P. Wolfe. An algorithm for quadratic programming. *Naval research logistics quarterly*, 3(1-2):95–110, 1956.

[GJ79]   M. R. Garey and D. S. Johnson. *Computers and Intractability: A Guide to the Theory of NP-Completeness (Series of Books in the Mathematical Sciences)*. W. H. Freeman & Co Ltd, New York, NY, 1979.

[GJS76]   M. R. Garey, D. S. Johnson, and L. Stockmeyer. Some simplified NP-complete graph problems. *Theoretical Computer Science*, 1(3):237–267, 1976.

[GLS88]   M. Grötschel, L. Lovász, and A. Schrijver. *Geometric Algorithms and Combinatorial Optimization*. Springer-Verlag, Berlin Heidelberg, 1988.

[GM07]   B. Gärtner and J. Matoušek. *Understanding and Using Linear Programming*. Universitext. Springer-Verlag, Berlin Heidelberg, 2007.

[Gro56]   A. Grothendieck. Resumé de la théorie métrique des produits tensoriels topologiques. *Bol. Soc. Mat. Sao Paulo*, 8:1–79, 1956.

[GRSW10]   V. Guruswami, P. Raghavendra, R. Saket, and Y. Wu. Bypassing the UGC for some optimal geometric inapproximability results. Electronic Colloquium on Computational Complexity, Report No. 177, 2010.

[GvL96]   G. H. Golub and C. F. van Loan. *Matrix Computations*. Johns Hopkins University Press, 3rd edition, 1996.

[GW95]   M. X. Goemans and D. P. Williamson. Improved approximation algorithms for maximum cut and satisfiability problems using semidefinite programming. *Journal of the ACM*, 42:1115–1145, 1995.

[Haz08]   E. Hazan. Sparse approximate solutions to semidefinite programs. In *Proceedings of the 8th Latin American conference on Theoretical Informatics (LNCS 4957)*, pages 306–316. Springer-Verlag, Berlin, 2008.

[HCA+10]   P. Harsha, M. Charikar, M. Andrews, S. Arora, S. Khot, D. Moshkovitz, L. Zhang, A. Aazami, D. Desai, I. Gorodezky, G. Jagannathan, A. S. Kulikov, D. J. Mir, A. Newman, A. Nikolov, D. Pritchard, and

G. Spencer. Limits of approximation algorithms: PCPs and unique games (DIMACS tutorial lecture notes). arXiv:1002.3864, 2010.

[Hel10] C. Helmberg. Semidefinite programming. Web site, http://www-user. tu-chemnitz.de/~helmberg/semidef.html, 2010.

[Hig91] N. J. Higham. Analysis of the Cholesky decomposition of a semi-definite matrix. In M. G. Cox and S. Hammarling, editors, *Reliable Numerical Computation*, pages 161–185. Oxford University Pressi, Oxford, 1991.

[HRVW96] C. Helmberg, R. Rendl, R. Vanderbei, and H. Wolkowicz. An interior-point method for semidefinite programming. *SIAM J. Opt.*, 6(2):342–461, 1996.

[Hås99] J. Håstad. Clique is hard to approximate within $n^{1-\varepsilon}$. *Acta Math.*, 182(1):105–142, 1999.

[Hås01] J. Håstad. Some optimal inapproximability results. *Journal of the ACM*, 48(4):798–859, 2001.

[Kal08] S. Kale. *Efficient Algorithms using the Multiplicative Weight Update Method*. PhD thesis, Princeton University, 2008.

[Kar99] H. Karloff. How good is the Goemans–Williamson MAX CUT algorithm? *SIAM J. Comput.*, 29(1):336–350, 1999.

[Kho02] S. Khot. On the power of unique 2-prover 1-round games. In *STOC'02: Proc. 34th ACM Symposium on Theory of Computing*, pages 767–775, 2002.

[Kho10a] S. Khot. Inapproximability of NP-complete problems, discrete Fourier analysis, and geometry. In *Proc. International Congress of Mathematicians, Hyderabad, India*, 2010. Available on-line at http://cs.nyu.edu/ ~khot/papers/icm-khot.pdf.

[Kho10b] S. Khot. On the Unique Games Conjecture. In *Proc. 25th IEEE Conference on Computational Complexity, Cambridge, MA*, 2010. Available on-line at http://cs.nyu.edu/~khot/papers/UGCSurvey.pdf.

[KKMO07] S. Khot, G. Kindler, E. Mossel, and R. O'Donnell. Optimal inapproximability results for MAX-CUT and other 2-variable CSPs? *SIAM J. Comput.*, 37(1):319–357, 2007.

[KL96] P. Klein and H. I. Lu. Efficient approximation algorithms for semidefinite programs arising from MAX CUT and COLORING. In *STOC'96: Proc. 28th Annual ACM Symposium on Theory of Computing*, pages 338–347. ACM, 1996.

[KLS00] S. Khanna, N. Linial, and S. Safra. On the hardness of approximating the chromatic number. *Combinatorica*, 20(3):393–415, 2000.

[KMS98] D. Karger, R. Motwani, and M. Sudan. Approximate graph coloring by semidefinite programming. *J. ACM*, 45(2):246–265, 1998.

[Knu94] D. E. Knuth. The sandwich theorem. *Electronic J. Combinatorics*, 1(1):A1,1–48, 1994.

[KO09] S. Khot and R. O'Donnell. SDP gaps and UGC-hardness for Max-Cut-Gain. *Theory of Computing*, 5(1):83–117, 2009.

[KP06] S. Khot and A. K. Ponnuswami. Better inapproximability results for maxclique, chromatic number and Min-3Lin-Deletion. In *Automata, languages and programming. 33rd international colloquium, ICALP 2006, LNCS 4051*, pages 226–237, Springer, Berlin, 2006.

[KSH97] M. Kojima, S. Shindoh, and S. Hara. Interior-point methods for the monotone semidefinite linear complementarity problem in symmetric matrices. *SIAM J. Opt.*, 7(1):86–125, 1997.

[KW92] J. Kuczyński and H. Woźniakowski. Estimating the largest eigenvalue by the power method and Lanczos algorithms with a random start. *SIAM J. Matrix Anal. Appl.*, 13(4):1094–1122, 1992.

[KZ97]   H. Karloff and U. Zwick.   A 7/8-approximation algorithm for MAX 3SAT? In *FOCS'97: Proc. 38th IEEE Symp. on Foundations of Computer Science*, pages 406–415, 1997.

[Las10]   J. B. Lasserre. *Moments, positive polynomials and their applications.* Imperial College Press, London, 2010.

[Lau03]   M. Laurent. A comparison of the Sherali–Adams, Lovás–Schrijver and Lasserre relaxations for 0-1 programming. *Math. Oper. Res.*, 28:470–496, 2003.

[LLZ02]   M. Lewin, D. Livnat, and U. Zwick. Improved rounding techniques for the MAX 2-SAT and MAX DI-CUT problems. In *Integer programming and combinatorial optimization. Proc. 9th international IPCO conference 2002. LNCS 2337*, pages 67–82, Springer, Berlin, 2002.

[Lov79]   L. Lovász. On the Shannon capcity of a graph. *IEEE Transactions on Information Theory*, IT-25(1):1–7, 1979.

[Lov03]   L. Lovász. Semidefinite programs and combinatorial optimization. In B. Reed and C. Linhares-Sales, editors, *Recent Advances in Algorithms and Combinatorics*, pages 137–194. Springer, New York, 2003. On-line at http://www.cs.elte.hu/~lovasz/semidef.ps.

[LR05]   M. Laurent and F. Rendl. Semidefinite programming and integer programming. In K. Aardal et al., editors, *Discrete optimization (Handbooks in Operations Research and Management Science 12)*, pages 393–514. Elsevier, Amsterdam, 2005.

[Mak08]   K. Makarychev. Quadratic forms on graphs and their applications. Ph.D. Thesis, Princeton University, 2008.

[Mat10]   J. Matoušek. *Geometric Discrepancy (An Illustrated Guide), 2nd printing.* Springer-Verlag, Berlin, 2010.

[Meg01]   A. Megretski. Relaxation of quadratic programs in operator theory and system analysis. In A. A. Borichev and N. K. Nikolski, editors, *Systems, Approximation, Singular Integral Operators, and Related Topics*, pages 365–392. Birkhäuser, Basel, 2001.

[MK87]   K. G. Murty and S. K. Kabadi.   Some NP-complete problems in quadratic and nonlinear programming. *Math. Programming*, 39(2):117–129, 1987.

[Mon97]   R. Monteiro. Primal-dual path-following algorithms for semidefinite programming. *SIAM J. Opt.*, 7(3):663–678, 1997.

[MR99]   S. Mahajan and H. Ramesh. Derandomizing approximation algorithms based on semidefinite programming. *SIAM J. Comput.*, 28(5):1641–1663, 1999.

[MS65]   T. .S. Motzkin and E. G. Straus. Maxima for graphs and a new proof of a theorem of Turán. *Canadian Journal of Mathematics*, 17:533–540, 1965.

[MT00]   R. Monteiro and M. Todd. Path-following methods. In H. Wolkowicz, R. Saigal, and L. Vandenberghe, editors, *Handbook of Semidefinite Programming*. Kluwer Academic Publishers, 2000.

[Nes98]   Yu. Nesterov. Semidefinite relaxation and nonconvex quadratic optimization. *Optim. Methods Softw.*, 9(1–3):141–160, 1998.

[NN90]   Yu. Nesterov and A. Nemirovski.   *Self-concordant functions and polynomial-time methods in convex programming.* USSR Academy of Sciences, Central Economic & Mathematic Institute, Moscow, 1990.

[NN94]   Yu. Nesterov and A. Nemirovski. *Interior Point Polynomial Methods in Convex Programming: Theory and Applications.* Society for Industrial and Applied Mathematics (SIAM), Philadelphia, PA, USA, 1994.

[NRT99]   A. Nemirovski, C. Roos, and T. Terlaky. On maximization of quadratic form over intersection of ellipsoids with common center. *Math. Program.*, 86(3):463–473, 1999.

[OSV79]   D. P. O'Leary, G. W. Stuart, and J. S. Vandergraft. Estimating the largest eigenvalue of a positive definite matrix. *Mathematics of Computation*, 33(148):1289–1292, 1979.

[OW08]   R. O'Donnell and Y. Wu. An optimal SDP algorithm for Max-Cut, and equally optimal Long Code tests. In *STOC'08: Proc. 40th Annual ACM Sympos. on Theory of Computing*, pages 335–344, 2008.

[Par06]   P. Parrilo. Algebraic techniques and semidefinite optimization. Lecture Notes, MIT OpenCourseWare, http://www.mit.edu/~parrilo/6256, 2006.

[PSU88]   A. L. Peressini, F. E. Sullivan, and J. J. Uhl. *The Mathematics of Nonlinear Programming*. Springer-Verlag, Berlin, 1988.

[Rag08]   P. Raghavendra. Optimal algorithms and inapproximability results for every CSP? In *STOC'08: Proc. 40th ACM Symposium on Theory of Computing*, pages 245–254, 2008.

[Rag09]   P. Raghavendra. Approximating NP-hard problems: Efficient algorithms and their limits. PhD Thesis, Department of Computer Science and Engineering, University of Washington, 2009.

[Rie74]   R. E. Rietz. A proof of the Grothendieck inequality. *Isr. J. Math.*, 19:271–276, 1974.

[RS09a]   P. Raghavendra and D. Steurer. Integrality gaps for strong SDP relaxations of unique games. In *FOCS'09: Proc. 50th IEEE Symposium on Foundations of Computer Science*, 2009.

[RS09b]   P. Raghavendra and D. Steurer. Towards computing the Grothendieck constant. In *SODA'09: Proc. 19th ACM-SIAM Symposium on Discrete Algorithms*, pages 525–534, 2009.

[Sha56]   C. Shannon. The zero-error capacity of a noisy channel. *IRE Transactions on Information Theory*, 2(3):9–19, 1956.

[Spe85]   J. Spencer. Six standard deviations suffice. *Trans. Amer. Math. Soc.*, 289:679–706, 1985.

[Spe87]   J. Spencer. *Ten Lectures on the Probabilistic Method*. CBMS-NSF. Society for Industrial and Applied Mathematics (SIAM), Philadelphia, PA, USA, 1987.

[Ste10]   D. Steurer. Fast SDP algorithms for constraint satisfaction problems. In *SODA'10: Proc. 21st Annual ACM-SIAM Symposium on Discrete Algorithms*, pages 684–697, 2010.

[Tul09]   M. Tulsiani. CSP gaps and reductions in the Lasserre hierarchy. In *STOC'09: Proc. 41st ACM Symposium on Theory of Computing*, pages 303–312, 2009.

[Val08]   F. Vallentin. Lecture notes: Semidefinite programs and harmonic analysis. arXiv:0809.2017, 2008.

[VB96]   L. Vandenberghe and S. Boyd. Semidefinite programming. *SIAM Rev.*, 38(1):49–95, 1996.

[WS11]   D. P. Williamson and D. B. Shmoys. *The Design of Approximation Algorithms*. Cambridge Univ. Press, Cambridge, 2011.

[WSV00]   H. Wolkowicz, R. Saigal, and L. Vandenberghe, editors. *Handbook of semidefinite programming. Theory, algorithms, and applications*. Kluwer Academic Publishers, Dordrecht, 2000.

[Ye04]   Y. Ye. Linear conic programming. Lecture Notes, Stanford University, available at http://www.stanford.edu/class/msande314/notes.shtml, 2004.

[Zuc06]   D. Zuckerman. Linear degree extractors and the inapproximability of max clique and chromatic number. In *STOC'06: Proc. 38th ACM Symposium on Theory of Computing*, pages 681–690, 2006.

Index

B. Gärtner and J. Matoušek, *Approximation Algorithms and Semidefinite*   245
Programming, DOI 10.1007/978-3-642-22015-9,
© Springer-Verlag Berlin Heidelberg 2012